S0-BCZ-297

Clocks in the Sky
The Story of Pulsars

Geoff McNamara

Clocks in the Sky

The Story of Pulsars

 Springer

Published in association with
Praxis Publishing
Chichester, UK

Mr Geoff McNamara
Science Teacher
Evatt
ACT
Australia

Cover image: John Rowe Animation www.JohnRowe.org

SPRINGER–PRAXIS BOOKS IN POPULAR ASTRONOMY
SUBJECT *ADVISORY EDITOR*: John Mason, B.Sc., M.Sc., Ph.D.

ISBN 978-0-387-76560-0 Springer Berlin Heidelberg New York

Springer is a part of Springer Science + Business Media (*springer.com*)

Library of Congress Control Number: 2008930770

Apart from any fair dealing for the purposes of research or private study, or criticism or review, as permitted under the Copyright, Designs and Patents Act 1988, this publication may only be reproduced, stored or transmitted, in any form or by any means, with the prior permission in writing of the publishers, or in the case of reprographic reproduction in accordance with the terms of licences issued by the Copyright Licensing Agency. Enquiries concerning reproduction outside those terms should be sent to the publishers.

© Copyright, 2008 Praxis Publishing Ltd.

The use of general descriptive names, registered names, trademarks, etc. in this publication does not imply, even in the absence of a specific statement, that such names are exempt from the relevant protective laws and regulations and therefore free for general use.

Cover design: Jim Wilkie
Editor: Laura Booth
Typesetting: BookEns Ltd, Royston, Herts., UK

Printed in Germany on acid-free paper

Contents

For Brenda and Amy,
The Two Brightest Stars
in My Life

List of Illustrations

Acknowledgments

This story has taken over two years to write. At all times I had in mind to write a story that was scientifically and historically correct. The reason is simple: I am a teacher and writing, to me, is an extension of that calling. As well as inspiring people to search for new knowledge, both teaching and writing call for adherence to the facts. While it is impossible to either write or teach without personal interpretation creeping in, it is important to make sure any opinions are based on legitimate information. Being an outsider to pulsar astronomy meant asking, sometimes repeatedly, 'Is this how it happened?' But the scientists I asked to check my work were as gracious as they are accomplished: almost without exception, the response of these tireless astronomers was meticulous and, above all, patient. On a personal note, I would like to thank all those scientists who offered encouraging remarks – sometimes unjustifiably complimentary – about my writing. It should be remembered, however, that this is a story about *their* remarkable achievements. I am a story teller. The scientists are the ones who have made, and continue to make, history.

The number of astronomers involved in pulsar astronomy is far too large to interview, and so I took the approach of contacting those directly involved in specific developments. I was fortunate enough to be able to contact many of the major players, including the original discoverers of pulsars. To all the people who added their insights and recollections, their understanding and their vision of pulsar astronomy, I extend my sincere thanks. If you are able to comprehend the wonder and fascination of pulsars, it is thanks to the scientists listed below, not only for making my writing accurate and honest, but for exploring the realm of pulsars.

The following have been instrumental in the completion of this book, and are listed in roughly the order of the chapters they either contributed to or checked. Roger Clay checked the early drafts of the first two chapters, prompting a major re-write of both. I would like to express sincere thanks to Bruce Slee, one of the pioneers of radio astronomy, for his recollections and corrections to the story of the development of radio astronomy.

The discovery of pulsars itself was a major historical event and gave rise to an entire field of astronomy. I have attempted to portray the events as they occurred back in 1967/68 as an unfolding story; I also attempted to have my account checked by those involved at the time. While the discovery is what is important, it is impossible to ignore the controversial implications for science in general of the events of the day. I therefore tried to present a balanced view with as little interpretation and opinion as possible, and then have the account checked for accuracy. Thanks are extended to Antony Hewish, Francis Graham-Smith, Franco

Pacini and John Gribbin who contributed their recollections of the time, and I am especially grateful to Professor Hewish for checking the final draft and making important corrections.

Alan Vaughan provided important information on the first pulsar surveys and checked the chapter on the early interpretation of pulsars. Dave Staelin and Richard Lovelace made important contributions to the story of the Crab pulsar and then checked the text, while Paul Ruffle and Ken Freeman also provided recollections and background information. Of special interest to me as a writer was the pre-discovery detection of pulsars by Jerry Fishman, as described in the chapter on the Crab Pulsar, and I am grateful to him for telling me his previously untold story and checking my interpretation.

The story of the detection of the first optical pulsar was tricky, since there was a number of conflicting recollections and strong opinions. Nonetheless, I have tried my best to synthesize the facts into a coherent story. I am grateful to the following for their contributions, patient advice and/or corrections to the chapter on the discovery of the first optical pulsar: Don Taylor, John Cocke, Mike Disney, Stephen Maran and Roger Lynds. In particular, John Cocke was kind enough to provide me with a CD recording of the actual discovery night, which makes fascinating listening and is a valuable piece of history. Brian Marsden was kind enough to clarify the details about the IAU circulars that announced the discovery of the optical pulsar. Special thanks to Roderick Willstrop who told me his fascinating story of the pre-discovery detection of the first optical pulsar.

Thanks go to Andrew Lyne for his many important contributions and corrections to the chapters on early pulsar surveys, pulsar planets and globular pulsars, and for checking my work. Russell Hulse and Joe Taylor both checked my interpretation of the discovery of the binary pulsar, for which I am grateful. Thanks go to Don Backer for checking the chapter on the millisecond pulsar discovery. Robert Duncan thoroughly checked the chapter on magnetars and again made important suggestions and corrections. I am grateful to Duncan Lorimer for telling me his story of the discovery of the first double pulsar, a story that tells as much about scientific protocol as it does about good science. Duncan Lorimer and Maura McLaughlin both contributed to and checked the chapters on the Parkes Multibeam Pulsar Survey and the discovery of Repeating Radio Transients (RRATs).

I would like to offer special thanks to Dr Richard Manchester of the Australia Telescope National Facility for telling me the story of the earliest pulsar surveys through to the incredible success of the Parkes Multibeam Pulsar Survey, for checking more chapters than he had time to, and for introducing me to pulsars all those years ago.

Finally, I extend thanks to Clive Horwood, my publisher, for being so encouraging and patient over what turned out to be a much longer project than anyone anticipated, and to John Mason for his many important suggestions and encouraging remarks.

Geoff McNamara
Canberra, Australia

List of Abbreviations and Acronyms

ASCA	Advanced Satellite for Cosmology and Astrophysics
AXPs	Anomalous X-ray Pulsars
BATSE	Burst and Transient Source Experiment
CAT	Computer of Average Transients
CGRO	Compton Gamma Ray Observatory
ETI	Extraterrestrial Intelligence
GRBs	Gamma Ray Bursters
HST	Hubble Space Telescope
IAU	International Astronomical Union
IPS	Interplanetary Scintillation
LIGO	Laser Interferometer Gravitational wave Observatory
LISA	Laser Interferometer Space Antenna
NOVA	NASA Open Volumes on Aerospace
NRAO	National Radio Astronomy Observatory
NRL	Naval Research Laboratory
RA	Right Ascension
RRATs	Repeating Radio Transients
RXTE	Rossi X-Ray Timing Explorer
SETI	Search for Extraterrestrial Intelligence
SGRs	Soft Gamma Ray Repeaters
SKA	Square Kilometer Array
SNR	Supernova Remnant
THEA	Thousand Element Array
VLA	Very Large Array

Prologue: Buried in the data

The enormous balloon drifted ever higher in the Texan sky. It was just after 6am on June 4th 1967. Hanging below the tenuous sphere was a new breed of telescope sensitive to the most powerful forms of radiation in the universe: gamma rays. Its goal was to look for gamma ray emissions from the Crab Nebula, the remains of a star that had long ago exploded as a supernova and witnessed here on Earth by Chinese and Japanese observers in 1054. Gamma ray astronomy was only young, and no one really knew what to expect. X-rays had been detected from this strange looking object, and it was anticipated that there was more to be found, hence the search for gamma ray emissions. As the telescope peered at the Crab, its observations were recorded to magnetic tape for later analysis.

What had become of the star that created the Crab? Was there anything left, as some theoreticians predicted? Could there be the charred remains of the original star? Could there be, as some thought, the as-yet unobserved 'neutron star' in the heart of the Crab Nebula? Students watching the balloon flight had been told by their supervisors to forget the idea: such musings were theoretical at best, and bore little resemblance to reality.

Yet something was sustaining the Crab Nebula, a tremendous powerhouse of energy that had kept it glowing for 900 years. Unknown to astronomers, a new breed of star deep within the Crab was beating out a fantastic rhythm. It wouldn't be the first of its kind to be found, nor would it be discovered by gamma ray astronomers. This strange object and its kin would first be detected at the other end of the electromagnetic spectrum by radio astronomers. But collectively they were destined to become one of the most significant discoveries in the history of science. In time they would become celestial laboratories for exploring everything from the structure of spacetime to the fundamental nature of matter. All that lay in the future, however. For now they remained unknown and undetected.

As the gamma ray telescope stared at the Crab, the nebula's heart imprinted its signature on the magnetic tape. Buried in the gamma ray data was the call of the Crab; tick, tick, tick...

Introduction

They're tiny, dense, and to look at, rather dull. They're not colorful, nor are they particularly photogenic. Conceptually they're difficult, and beyond a rudimentary understanding most people probably shrug them off as being an astronomical also-ran and go back to reading about the Big Bang. They're pulsars. Dense spheres rotating at unbelievable speeds spreading radio beams out into space like cosmic garden sprinklers. Who would have thought that these unassuming objects would have such a tremendous impact on our understanding of the Universe. This is not mere hyperbole either. When astronomers had to justify building the world's largest radio telescope, the Square Kilometre Array (SKA) – an instrument so large and expensive that it is taking a consortium of nations to foot the bill – they cited pulsars as being one of the five main reasons. Other reasons for the SKA include the search for the origins of the Universe, and detecting Earth-like planets elsewhere in the Galaxy. Pulsars, it turns out, have the potential to reveal such diverse aspects of the Universe as the nature of spacetime and the fundamental properties of matter. In short, pulsars are a treasure of knowledge about the physical universe, and those that study them are unsung heroes of modern science.

Pulsars are bizarre, violent and immensely energetic objects, fitting tributes to their origins in supernova explosions, which themselves mark the violent ends of once brilliant suns. Originally the size and mass of a star, each pulsar is reduced by gravity to a sphere the size of a city: the once tenuous stellar gas compressed to a density so great that a handful of the stuff would weigh a million tonnes. At such density, a pulsar's gravity is so great that it wraps space and time around itself like a blanket. This immense yet compressed bulk spins with unimaginable ferocity creating waves in the fabric of spacetime that ripple through the Universe. And as they spin they emit beams of radiation – light, radio and X-rays – that stream out into space. If one of the beams crosses Earth the pulsar seems to flash on and off like a warning beacon marking the site of a stellar cataclysm.

Such a range of unique characteristics makes pulsars laboratories for testing phenomena with not just astronomical but cosmological significance. The beams of radiation emitted by pulsars cross vast expanses of space providing astronomers with probes of the interstellar environment. The first extrasolar planets – planets outside of our Solar System – were discovered orbiting pulsars, the last place anyone expected to look. The incredible accuracy of pulsars allows astronomers to study one of the most important predictions of general relativity, gravitational waves. Pulsar rotation rates are so precise they are regarded as among the most accurate timing sources known. Every period is precise and every pulsar unique so that although still in the realm of science fiction, pulsars might one day serve as navigational beacons for interstellar spacecraft. From the cosmological to the planetary, pulsars have continued to keep thousands of scientists busy for decades.

Pulsars were an unexpected find that changed astronomy forever. Predicted long before their discovery, their identification was the result of the unlikely convergence of two entirely different branches of science: nuclear physics and radio astronomy. While one descended into the nucleus of the atom searching for fundamental particles and states of matter, the other began trawling the skies looking for recently discovered, intense sources of radio waves far beyond our Milky Way galaxy. As the story unfolds we will look over the shoulders of some of the world's finest scientists as they develop new machines and new ideas. We will be there as the first pulsar is discovered on a winter's night in 1967; sit in on the first announcement of their discovery following months of unprecedented scientific secrecy; witness the discovery of the first millisecond pulsar as it spins hundreds of times each second; and stand and watch as astronomers detect the first binary pulsar. To say the least, pulsars are full of surprises, and this may never be better illustrated than the fact that the first planets beyond our Solar System were found orbiting these stellar corpses. After all this we will look to the future to see what may be: the potential discoveries of a radio telescope with a surface area of a city block, probes into the very nature of existence.

You cannot understand the story of pulsars without understanding the three converging branches of science that led to their discovery. First there was the emerging understanding of the nature of stars in the late 19th and early 20th centuries. The 1930s saw two events that were to converge three decades later: the discovery of the neutron, and the emergence of radio astronomy. The story of pulsars is much fuller for knowing about all three developments. The discovery of pulsars actually took place in discrete episodes: they were predicted, then detected, then recognized. At the time of their prediction it was thought neutron stars would be completely invisible and hence no one bothered to look for them. Four decades on the search for pulsars in the Galaxy and research into their physics has involved hundreds of careers, thousands of research papers and two Nobel Prizes. And yet the way they power the intense beams of radiation that announce their existence is still not well understood. The mechanics, the precision timing, the visibility and the appearance of pulsars has all been documented. But what makes pulsars tick is still somewhat of a mystery.

This is the story of pulsars, of their stunning prediction from the same physics that led to the threat of nuclear war, their unexpected discovery and the controversies that followed, what astronomers have learned about them in four decades and what they hope to learn about them in the future. It is a story of science and scientists, of great advances in instrumentation and ingenuity, of one of the most powerful astrophysical phenomena known which is able to probe one of the most subtle and far-reaching. But the history of pulsar science is also littered with controversy, threatened careers, careless comments to the media, the balance between open scientific communication and the rivalry that hides within and between institutions. It is a story of international cooperation and cutting edge research in technology. Above all, however, it is a story of fantastic science.

Along the way we'll meet some of the finest minds in the history of astronomy, some more widely acknowledged than others. Of course it's impossible to paint an adequate picture of all of the players, but for the key scientists I have included a biographical paragraph or two. These are the people who made the pulsar story, and the story is incomplete without knowing a little of them as people and as scientists. One of the world's leading pulsar researchers once remarked to me that what keeps them interested in pulsars for an entire career is that every few years something totally unexpected shows up. The best example of this was the discovery of pulsars itself. The initial observations were so unexpected, so inexplicable, that the surprise and caution of the astronomers involved caused them to delay announcement of the results in case they had made a mistake. They hadn't, and since then the story of pulsar science has unfolded in ways just as unexpected as their initial discovery and it is likely that this trend will continue. The most recent example was the discovery of a new type of pulsar, unceremoniously dubbed RRATs, that may far outnumber the 'conventional' pulsars. It's been like that all along: no sooner do astronomers feel they are getting close to an understanding of pulsars than nature, or rather technology, reveals something entirely new. And as I said at the beginning of this introduction, astronomers are convinced there's more to discover. Much more. This is the story so far.

1 'Life & Death Among The Stars'

'It is not too much to hope that in not a too distant future we shall be able to understand so simple a thing as a star.'
Arthur Stanley Eddington, 'The Internal Constitution of the Stars'

The Nature of Stars

To understand pulsars we need to understand stars. Pulsars emerge from the ashes of stars that have perished in spectacular explosions called supernovae and it is with these titanic events that this chapter will close. But before we go there, it is important to understand at least the basics of what makes stars shine, where they get their energy from, why they last so long and why they don't last forever. Also fundamental to our story is why it is that not all stars produce pulsars: why it is that only the most massive stars produce them, and only in death. Our Sun is a hundred times the diameter of Earth, a gargantuan sphere of plasma that has and will produce enormous amounts of energy for billions of years. Despite the Sun's immense dimensions, it is considered a dwarf among stars. The Sun will never produce a pulsar: it is simply too small. To produce a pulsar calls for a massive star many times larger than the Sun.

We now know that stars are huge spheres of gas, mainly hydrogen and helium. Gravity not only holds the gas of a star together in a sphere, but is forever trying to collapse it into an ever smaller volume. This compresses the material in the center of the star. So great is the resulting heat *and* pressure that atoms of lighter elements are fused together to produce heavier elements, releasing energy in the process. As that energy is released it prevents the star from collapsing in on itself entirely. As long as there is sufficient fuel, the energy released within the star will keep the star inflated against the pull of gravity. But when the fuel runs out, as it inevitably must, the star will collapse on itself and perish. This is the ultimate fate of all stars.

So, stars may seem eternal, but that is an illusion. They are born, they live, and they die. How do we know? Take the Sun, for example. Our species was not around to see the origin of the Sun, and will be long gone by the time the Sun comes to an end. What we see in the daytime sky is a mere snapshot of a middle aged star, and during the entire history of humanity the Sun will change hardly

Figure 1a Spectra of different kinds of stars. Against the normal background of the spectrum, specific lines can be seen which identify the type of star. (Courtesy European Space Agency.)

at all. We do not see it age. But there are other ways of finding out about the Sun's past and future. By looking out into space we can see not only that the stars are other suns, we can infer from observations of stars at different phases of their lives how the Sun formed and what is its likely end. It's like standing in the middle of a crowd of people of different ages, from birth to old age. Being a person like those we see around us, we can infer that what has happened to them has and will happen to us. The same approach has been applied to understanding stars.

There's a major problem when it comes to studying the stars scientifically, however: stars are too far away to touch. All of the sciences have at their heart the notion of experimentation with the goal of testing hypotheses. In any experiment the goal is to modify a variable and see how the results change. Astrophysicists can't do this: the stars are just too far away and far too vast in terms of size and energy for us to interfere with. The solution is to let nature perform her own experiments, of which there is a seemingly infinite range, and look for patterns in the collected results. In addition, astronomical observation depends on and frequently determines the development of new instruments. New discoveries have both depended on and driven the development of not only larger and more sophisticated telescopes but also the sensitivity and diversity of the instruments that utilize the light they gather from the distant stars. Arguably the most important breakthrough in astronomical instrumentation following the invention of the telescope itself came in the early 1800s with the development of spectroscopy.

The Rise of Spectroscopy

Spectroscopy is based on the fact that light can be broken up into its various colors, a phenomenon known as dispersion. A rainbow is a natural example of dispersion. Dispersion was investigated by Newton a century and a half before the development of spectroscopy. In his famous experiment, Newton allowed sunlight to pass through a small hole in a window shutter and through a glass prism. He saw what many of us are familiar with: a beautiful array of colors shining as brightly as the original sunlight. However, Newton took the experiment a step further. He allowed the colored light to pass through a second prism and reproduced white light. White light was not a pure entity, but a mixture of all the colors of the rainbow.

What Newton didn't see was a series of dark lines within the spectrum, as if thin slices of light had been surgically removed. Although the British physicist William Hyde Wollaston did notice them in 1802, the first detailed study of the lines was made by an outstanding German optician, Joseph von Fraunhofer. Not only did Fraunhofer use his superior instruments to map hundreds of these dark lines, lines which now bear his name, but he applied the technique to the Moon and planets. He saw that the arrangement of lines in planetary spectra was unsurprisingly the same as the Sun's, whose light they reflected. But when he observed the spectra of the stars he saw that the arrangement of the lines in stellar spectra were quite unlike that found in the spectrum of the Sun. Here was a stellar equivalent of a bar code detailing the identity and make-up of the stars. But Fraunhofer was an optician, not a physicist, and he died not knowing the importance of what he had seen.

The identification of spectral lines started astronomers thinking about the classification of stars. This grew initially out of the work of the American amateur astronomer Lewis M. Rutherford and the Italian Jesuit Angelo Secchi. But the real work of classifying stars based on the arrangement of spectral lines was carried out by a team of women working under the astronomer Edward Charles Pickering at Harvard College Observatory. Not permitted to enroll in the courses delivered by the university that employed them, over a dozen women analyzed the spectra of thousands of stars and began classifying them accordingly. By this stage it was known which lines were produced by at least a few of the elements, and so at first the stars were classified according to the strengths of the lines due to hydrogen, allocating each star a letter A, B, C, and so on. But one of the analysts, Annie Jump Cannon, realized that there was a progressive change in the appearance of the spectra and so rearranged the sequence. A heritage of this is the jumbled notation O B A F G K M R N, which professional and amateur astronomers learn today. Now this was pure taxonomy: no one knew what the classifications meant. It was clear that the spectral lines indicated something different about the stars, perhaps even a relationship. What was needed was a physical explanation of what that relationship was. The answer emerged out of the work of a string of scientists, work that eventually united two previously unconnected branches of science: astronomy and nuclear physics.

The Hertzsprung–Russell Diagram

Fraunhofer's lines indicated the chemical composition of different substances and it was thought that the same would be true of the stars. To a certain extent this is true, but as surveys of stars began to accumulate the idea became less and less satisfactory. In the early 1900s, two astronomers working independently made the crucial connection that was to pave the way to a true understanding of the evolution of stars. Henry Norris Russell working at Princeton and a Danish chemical engineer-turned-astronomer, Ejnar Hertzsprung both carried out analyses of spectra of stars in two large, well-defined clusters of stars called the Hyades and Pleiades. Visible to the naked eye in the early evening skies of December, these vast collections of stars are obviously physically associated in space and hence lie at the same distance from Earth. Now the apparent brightness of a star in the night sky depends not only on its intrinsic brightness but also how far away it is. Being at the same distance meant that any differences in the observed properties between the stars in either the Pleiades and Hyades stars clusters – such as how bright the stars appear – must be due to the intrinsic properties of the stars themselves.

Astronomers already knew how to measure the distances to the stars using a technique known as parallax. This phenomenon is familiar to anyone who has watched the relative movement of nearer and farther objects from a moving car or train: nearer objects move much more obviously against the more distant background. The same can be seen among the stars. As Earth moves from one side of its orbit to the other, the nearer stars appear to move against the more distant backdrop of stars. Careful measurement of this relative change in position allows astronomers to determine the distances to the nearer stars. Comparing the apparent brightnesses of stars with their distances determined from parallax measurements yields the true brightnesses of the stars, and with them a host of other physical properties, not the least of which was how much energy they produce. Stars at the same distances, which is the case for stars in the Pleiades and Hyades clusters, meant those stars could be compared directly with one another in terms of energy output, in other words, their intrinsic brightness. From this reasoning and careful observation, Hertzsprung and Russell independently showed that the different line widths in the stellar spectra were directly related to the intrinsic brightnesses (i.e. luminosities) of the stars.

There was one more implication of stars that were physically associated in space: they must all have formed from the same primordial material and therefore all be made of the same stuff. That meant differences in the spectra of stars in the same cluster couldn't be explained by differences in chemical

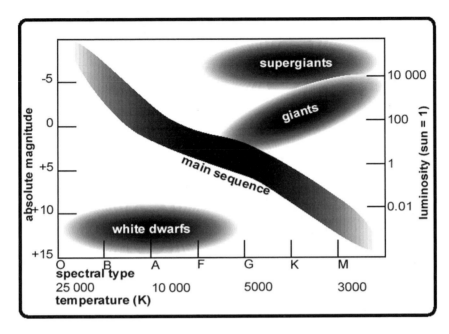

Figure 1b The Hertzsprung–Russell diagram compares the brightness and temperature of stars. The majority of visible stars are found along the main sequence. As a star ages, its location on the H–R diagram changes, tracing out a path. The path passes through different known stages of stellar evolution, such as the giants and white dwarfs. Illustration by the author.

composition alone. Astronomers were by this stage able to determine the temperatures of the stars by looking at star colors: just as a piece of metal in a furnace glows different colors as it heats up, so the colors of the stars are an indication of their surface temperature. But the brightness of a star is not just a matter of how hot it is: 'cool' stars[1] can be brighter than hot stars simply by being bigger. For stars of different brightness but the same temperature, luminosity must depend on physical size. Now the connections were all in place: by comparing the brightness of a star, its intrinsic luminosity measured from the widths of the spectral lines, and the temperature as determined from its color, astronomers for the first time were able to find the distances to the stars not by parallax, but by spectroscopy.

By plotting the spectral type (color/temperature) of stars against their brightness, Russell and Hertzsprung produced a simple chart consisting of a series of gentle curves made up of the stars they had plotted. A quick glance at the Hertzsprung–Russell (H–R) diagram strongly suggests more than a casual relationship between the stars. For a start, the vast majority of stars lie along a band that runs from the upper left corner (bright and hot) to the lower right (dim and cool), a collection known as the main sequence. Yet there were other collections of stars, such as bright yet cool stars in the upper right, and dim, hot stars scattered along the bottom. The more stars astronomers plotted, the more it seemed the curves represented evolutionary changes in the stars. To return to our analogy of standing in a crowd of people trying to infer a single human life cycle, it was the equivalent of plotting two characteristics of humans, say height and wrinkliness, on a chart and noticing that only tall people had wrinkles, but not all tall people. Clearly there is a relationship. Astronomers were convinced that there was an evolutionary relationship represented in the H–R diagram. Both Hertzsprung and Russell offered explanations of what the relationship was. Both were completely wrong.

Development of Stellar Theory

The explanation wouldn't become clear until astronomers understood what was going on deep within the stars. It was clear that stars were immense in size, produced copious amounts of energy and did so for a very long time. But where was all the energy coming from? In the first decade of the 20th century Arthur Schuster and Karl Schwarzschild had created a model of the Sun in which temperature and density increase with depth. In this way, it was shown that the inward pull of gravity could be balanced by the outward flow of radiation. Arthur Eddington started with this theory but extended it, and in his classic work *The Internal Constitution of the Stars*, he set down his theory of the structure of stars. From his calculations, Eddington was able to introduce time into his theory and showed that stars on the H–R diagram represented an evolution from one type to another. What was extraordinary was the amount of time that evolution took, measured no longer in millions of years, but in billions. What could keep the Sun and other stars shining for so long?

The answer emerged from two new ideas in science: atomic physics and the theory of relativity. Up until then the only viable source of a star's energy was gravitational collapse, but around the turn of the century a new type of science was emerging that would explain not only the age of the Earth and Sun, but the power of the stars. Marie and Pierre Curie discovered that certain elements can transform into lighter elements through radioactive decay. Uranium, for example, decays over time into the lighter, more stable element lead, with some of the mass of the original atom being converted to energy. This phenomenon is known as nuclear fission. By the time Eddington tackled the problem of the nature of stars, radioactivity was accepted as a long-term source of energy. At one stage it was even suggested that the Sun could be powered by the decay of radium: it would only take a tiny amount, just 3.6 grams per cubic meter of the Sun. Although on the wrong track, at least alternative sources of energy were being discussed. The breakthrough came when it was realized that not only the fission of heavy elements, but also the fusion of lighter elements into heavier elements, would release enormous amounts of energy.

Both hydrogen and helium had been identified in the spectra of stars. It had been shown that the mass of helium nucleus was a tiny bit less (0.5%) than the

COMPUTERS PLUS+ of Kennesaw Georgia

Computer/Laptop Repair and Sales of Kennesaw Georgia

TOSHIBA hp ASUS

Mon- Friday 10am-8pm
Saturday 11am-6pm
Sunday 12pm-4pm

Ph: 770-693-0769
computersplusga@gmail.com
3060 Cobb Pkwy NW #105
Kennesaw, GA 30152
(Next to Oyster Cafe)

Computerrepairofkennesaw.com

mass of the four hydrogen nuclei from which
hydrogen into a helium release sufficient energy
be destroyed, but it can be converted into energy
famous equation $E = mc^2$. Even though the diffe
hydrogen and helium atoms is small, when multiplie
of light even 0.5% converts to an enormous amount o.
getting hydrogen nuclei to fuse. Eddington calcu.
temperature of the Sun was around 40 million degrees;
produce nuclear fusion? A hydrogen atom is simply a pos.
with an electron, and at such high temperatures the protons
sort of particle soup with the protons free to interact. Like c.
fusing two positively charged protons means overcoming this   ll,
known as the Coulomb barrier. The solution lay in the fact tha   ..crostatic
force holding individual protons apart isn't the only force they respond to. The
Universe is governed by four fundamental forces: electromagnetism (the force
holding protons apart), gravity, the weak nuclear force and the strong nuclear
force. It's the last, much stronger than electromagnetism, that holds protons
together in the nuclei of atoms. But it only does so over a very short distance,
beyond which electromagnetism takes over.

Just how you get protons close enough for the strong nuclear force to
dominate was discovered in 1929 by the Russian born American physicist
George Gamow. The process he came up with was called 'quantum tunneling'
and was a kind of back door into the atomic nucleus. It allowed protons to bond
with each other to form newer, heavier elements, giving off energy in the
process. Now this kind of fusion doesn't occur at room temperature: the fact
you are sitting on a chair rather than fusing with it or falling through it is due
to electrostatic repulsion of the electrons in the atoms that make up you
repelling the electrons in the atoms of the chair. But in the unimaginable
temperatures and pressures inside something so massive as a star, things are
different. There the temperature is so high, and the atoms moving with such
tremendous speeds, that they occasionally slam into one another with enough
energy to overcome electrostatic repulsion and fuse into new, heavier particles
and ultimately elements. Now this doesn't happen very often, but because
there are so very many atoms inside a star even this low probability means that
enough collisions take place to power the star. Furthermore, because stars are so
very large, the energy release would continue for a very long time, measured in
billions of years. Eddington's prediction of atomic energy being the lifeblood of
stars had been fulfilled at last.

Nucleosynthesis

We owe a lot to the fusion of elements inside stars; after all, we're made of the
same stuff. The details of how stars turn simple elements into heavier ones – a
process known as nucleosynthesis – was worked out in the late 1950s by a team

of scientists Fred Hoyle, William Fowler, Geoffrey Burbidge and E. Margaret Burbidge. As you can imagine, elements with smaller numbers of protons are 'easier' to fuse together simply because there is less repulsion to overcome. The smallest element is hydrogen with a nuclear charge and mass of just 1. Nonetheless the feat requires tremendous temperatures: the fusion of hydrogen at a temperature of about 15 million K.[2] The product of the union of hydrogen nuclei is helium with a nuclear mass of 4. Being heavier, helium sinks to the center of the star, so the hydrogen forms a thick layer above it like the white of an egg surrounding the yolk. The hydrogen fusion continues in a shell around the helium core, keeping the star shining as brightly as before. What happens next depends on the initial mass of the star, and as we will see, the fates of different mass stars are extraordinarily different.

The End of the Sun

In the case of the Sun the temperature and pressure builds up inside this new helium core so that at a temperature of 100 million K the helium itself begins to fuse to form carbon. Being heavier, the carbon falls 'beneath' the helium to the center of the star, where the pressure and temperature continue to rise relentlessly. At this stage the Sun's outward appearance will change dramatically. Swelling to more than 90 times its main sequence size and engulfing the planets Mercury and Venus,[3] a star the mass of the Sun swells and cools. As the Sun increases in size it grows brighter, but the cooling of the outer atmosphere causes its color to turn reddish: it has become a red giant. On the H–R diagram it moves from the main sequence to the upper right. All the while, deep inside the star remains its hot, dense core.

Becoming a red giant spells the end of life for a star of less than eight solar masses. Such stars are unable to generate the temperatures needed to fuse carbon to oxygen because it is cooled by the continuous escape of subatomic particles called neutrinos. These bizarre particles pass from the center outward through the star without interacting with matter carrying energy from the star. The core, now made of carbon and oxygen, is surrounded by a tenuous sphere of hydrogen and helium. Stars with a mass of between eight and ten times the mass of the Sun are sufficiently massive to generate the temperatures and pressures needed to burn carbon to oxygen, neon and magnesium. After an amount of time that depends on its initial mass, the star resembles an onion, with layer upon layer of elements of increasing weight from the outermost hydrogen through helium, carbon, and so on all the way to the core. But this is not enough to prolong the life of the star, and in time the star runs out of nuclear fuel. Radiation pressure from the core gently pushes the shell of gas away from the core over hundreds of thousands of years. The gently glowing shell becomes a planetary nebula, and at the center is revealed the core, now regarded as a 'white dwarf' star, a slowly cooling ember about the size of the Earth. Seen from the outside, the star appears to increase in temperature and yet because it is so small compared with main sequence stars its brightness falls. Plotted on the H–R diagram, the star moves from the region of the red giants in the upper right, downward and to the left. Its track passes through the main sequence and finally ends in the lower left of the H–R diagram, languishing for billions of years as it slowly cools and fades from the realm of the stars.

Evolution of High Mass Stars

Of greater importance to the story of pulsars is the fate of more massive stars. If the initial mass of the star is more than ten times the mass of the Sun it is able to exert sufficient pressure and heat to continue burning successively heavier elements. As it does so, it glows far brighter in space than its low mass cousins and at a far greater pace: the greater mass of the star means more of the crushing weight at the core, resulting in a furious rate of nucleosynthesis. Where a star the mass of the Sun will last perhaps 8 billion years, a massive star will last 100 million years. During its short life, a massive star continues to make heavier elements, piling one upon the other in the core: the heavier elements sink to the center while the successive layers of lighter elements surround it like the layers of an onion. Its demise is brought about by the build up of iron in the core. Rather than releasing energy, the fusion of iron nuclei *requires* energy. Until now, the fusion of lighter elements into heavier elements has been a source of energy for the star, energy that has kept it inflated against the indefatigable pull of gravity. But gravity – the weakest yet farthest reaching of all forces in the Universe – wins in the end. Having used up its supply of nuclear fuel, the star has nothing left to keep it inflated against its own weight. The nuclear burning processes that have kept it alive for its short life now shut down. Over a tiny amount of time, the bulk of the star's outer layers collapse inward, rushing toward the core. There they collide in a titanic impact so great that in an instant elements heavier than iron – elements that make up our familiar worlds – are forged. The tremendous implosion results in a 'core bounce': the fusion of matter around the core at the time of collapse releases vast amounts of energy. The star explodes brilliantly, hurling the newly formed elements in all directions. The eruption blazes with the light of a hundred billion stars. It has become a supernova. For a brief time it outshines the entire Galaxy.

As the outer layers of the star surge outward into space, the stellar core continues to collapse. With the removal of radiation pressure, gravity takes over and crushes the core to unimaginable densities. From the death of the massive star, a new strange object emerges: glowing brightly and rotating at fantastic speed, so dense it warps spacetime around it and begins radiating intense beams of radio waves into space.

References

1. 'Cool' is a relative term here. The temperatures of the coolest stars are still measured in thousands of degrees.
2. K is shorthand for Kelvin, the equivalent of degrees Celsius measured from absolute zero.
3. Possibly Earth as well, but in terms of human heritage it is a moot point: either way the entire surface of our blue green world will have been melted beyond recognition.

2   '1932'

Chadwick's Identification of the Neutron

In January 1932, a young British researcher, James Chadwick, read a paper published in the journal of the *French Academy of Sciences*, Comptes Rendus by a husband-and-wife team, Frederick and Irene[1] Joliot-Curie. In that paper, they described how they had bombarded a piece of the gray, toxic metal beryllium with α (alpha) particles[2] shot out of a sample of the radioactive element polonium. What emerged from the experiment was a highly energetic yet intangible form of radiation. Although they couldn't detect the radiation directly, they were able to detect protons punched out of a range of materials placed in the path of the new radiation. What puzzled them was this new form of radiation passed through almost anything more freely than any other form of radiation they had previously encountered. Earlier experiments by Walter Bothe and his student Herbert Becker in Germany had shown that the radiation would pass through 200mm of lead. Not bad, considering it takes less than 1 mm of lead to stop the most energetic form of energy then known, γ (gamma) rays.[3] This new radiation was 'so hard that one can hardly doubt their nuclear origin'.[4] Both teams assumed that the unknown radiation must in fact be an even more energetic form of gamma rays; after all, it was known that gamma rays could knock electrons out of metals, so why not protons out of the nuclei of other materials? But this was only part of the story; Chadwick knew the rest. He knew what they had found but had not identified. Electrified into action, he repeated and extended the experiments and, in the course of three weeks of frantic, intense experimentation, demonstrated the existence of the third component of the atom, the neutron. The discovery was to have profound consequences for events in the heavens and here on Earth.

James Chadwick was born in Manchester, England on October 20th 1891. After graduating from Honours School of Physics in 1911 he spent the next two years working under the New Zealander Ernest Rutherford in the Physical Laboratory at Manchester University. Owing to financial problems, Chadwick moved to Berlin to study at Physikalisch Technische Reichsanstalt at Charlottenburg under Professor Hans Geiger (of Geiger counter fame). The following year, war broke out. The travel company, Thomas Cook, advised British citizens that they were in no immediate danger and therefore there was no need to flee Germany, and yet it wasn't long before Chadwick was detained as a civilian POW in the disused stables of a racetrack at Zivilgefangenenlager, Ruhleben where he spent the next four years. Although he was denied the opportunity to conduct experiments, he was permitted to read and to talk with other physicists. Despite the repressive situation, Chadwick did manage to conduct limited research using a popular toothpaste being marketed at the time by Berlin Auer. The toothpaste contained thorium which was intended to make teeth glow brighter. ('Use toothpaste with thorium! Have sparkling, brilliant teeth – radioactive brilliance!')

At the end of the war, Chadwick returned to Manchester where he worked with Rutherford once more on the transmutation of elements, the changing of one element into another. Alchemists had tried to accomplish this feat for centuries but were utterly unsuccessful. Eddington had suggested that the stars were able to do what alchemists could not; now Rutherford wanted to know how. At the end of the 19th century, Rutherford had carried out experiments that had shown the existence of the nucleus of the atom. By firing particles at gold foil, he had shown that while most flew straight through, some were reflected. Rutherford famously exclaimed that it was 'as if you fired a 15-inch naval shell at a piece of tissue paper and the shell came right back and hit you.' This experiment convinced Rutherford that the atom was composed mainly of nothing save a phantom-like cloud of electrons surrounding a tiny, massive and impenetrable nucleus.

But what of the nucleus? What was it made of? At the time there were two particles known: the negatively charged electron and the positively charged proton, with protons by far the heavier of the two. The problem was that the masses of different elements just didn't add up. A hydrogen nucleus had a charge of +1 and logically weighed as much as a proton. A helium nucleus, on the other hand, had a charge of +2 but was four times as massive as the hydrogen nucleus. To be this heavy it must have had four protons in the nucleus giving a charge of +4, but to balance the electrical charge there had to be two electrons in there as well. Electrons aren't weightless, but as far as scientists were concerned they were so light compared with protons as to make little difference to the mass, at least considering the problems they solved.

There was another problem with this model, however. In 1925 two Dutch scientists, George Uhlenbeck and Samuel Goudsmit, identified a fundamental characteristic of nuclei, protons and electrons called 'spin', with both particles having a spin of either $+\frac{1}{2}$ or $-\frac{1}{2}$, with no in betweens. Now when you add up the particle spins of an odd number of particles, the result will be one of half integers, say $1\frac{1}{2}$ or $3\frac{1}{2}$. But when the total spin of a nitrogen nucleus was measured it was found to be a whole number, despite having a total of 21 particles: 14 protons and 7 electrons to counter the positive electrical charge of 7 of the protons in the nucleus, as per Rutherford's theory. There had to be a third type of particle lurking deep within the atom that had nothing to contribute to its charge but increased the mass of the atom by as much as a proton. Rutherford speculated that this heavy, neutral particle must be a combination of the massive proton and light-weight electron which permitted nuclei to have the masses observed. In his Bakerian lecture to the Royal Society in 1920, Rutherford explained that 'it may be possible for an electron to combine much more closely with the [hydrogen] nucleus, forming a kind of neutral doublet.' Rutherford named this hypothetical particle the neutron.[5]

Rutherford and his students spent the next twelve years searching for the neutron but without success. In 1964 Chadwick recalled that some of these attempts were 'so desperate, so far fetched as to belong to the days of alchemy'.[6] But as soon as Chadwick read the Joliot-Curie paper, he knew he was about to find this elusive particle. As discussed at the start of this chapter, the Joliot-Curies had suggested that the penetrating radiation consisted of gamma rays. Chadwick rejected this explanation pointing out that gamma rays were not energetic enough to produce the observed results: they simply lacked the energy needed to punch protons from the different materials tested at the observed velocities. Three weeks after the publication of their results, the Joliot-Curies had changed their mind on the matter also, but by then Chadwick had beaten them to the finish line.

Chadwick wasted no time in setting up a replica of the Joliot-Curie experiment. Despite Rutherford's rule of no one working in the laboratory past 6pm, being unwilling to waste time or take a risk that his delicate experimental equipment may be disturbed by other researchers, Chadwick continued late into the night. Chadwick also used polonium as a source of α (alpha) particles. In front of this he placed a sample of beryllium from which came the mysterious penetrating rays. In the path of these rays he placed targets of various kinds: hydrogen, helium, lithium, beryllium, carbon, air and argon. Chadwick then placed a detector in front of the samples in order to detect any particles coming from them. The particles he detected were protons being knocked out of the samples by the unknown radiation. By measuring the speed with which the particles were knocked out of the samples, Chadwick was able to show that gamma rays just weren't energetic enough to do the trick. The penetrating radiation had to consist of particles with at least the mass of a proton; and yet they had no electrical charge. As he wrote in a letter to the journal *Nature* on February 27th 1932, 'The difficulties disappear, however, if it be assumed that the radiation consists of particles of mass 1 and charge 0, or neutrons.'

Exhausted but exhilarated, Chadwick announced his discovery at an informal gathering of physicists known as the Kapitza Club.[7] The response of those present was enthusiastic and acknowledged the importance of the discovery. In a more detailed account of his experiment published in *The Proceedings of the Royal Society*, Chadwick expressed his discovery a little more cautiously: 'It is evident that we must either relinquish the application of the conservation of energy and momentum in these collisions or adopt another hypothesis about the nature of the radiation.' These days the law of conservation of energy – that energy can be neither created nor destroyed – is taken for granted. Back in 1932, however, there was still debate over whether the same conservation laws related to the quantum world as they do in the macroscopic world. Nonetheless, Chadwick continued,

'If we suppose that the radiation is not a quantum radiation, but consists of particles of mass very nearly equal to that of the proton, all the difficulties connected with the collisions disappear, both with regard to their frequency and to the energy transfer to different masses. In order to explain the great penetrating power of the radiation we must further assume that the particle has no net charge. We may suppose it to consist of a proton and an electron in close combination, the 'neutron' discussed by Rutherford in his Bakerian Lecture of 1920.'

And what of the Joliot-Curies? Frederick was understandably disappointed at being scooped by Chadwick, and privately expressed his frustration with the practice of having to publish scientific evidence in order to gain priority while watching other researchers build on that published work. After all, this was what spurred Chadwick on to success. Nonetheless a gracious Frederick publicly gave Chadwick credit for the discovery.

Implications

Being electrically neutral, neutrons are completely unaffected by the positive repulsive force of protons in the nucleus of an atom. They therefore have a remarkable ability to penetrate the nuclei of other atoms. In early 1939, Otto Hahn working in Berlin had shown how the uranium nucleus, when bombarded by neutrons, split apart and released enormous amounts of energy. Two months later the Joliot-Curies published a letter in *Nature* announcing that not only did the uranium nucleus split apart, but that along with energy, it released yet more neutrons which would invade and split further uranium nuclei releasing still more energy. The concept of a chain reaction had been realized, and the consequences were thrilling to some, terrifying to others. Physicists of both persuasions realized that the Joliot-Curie results pointed the way to the release of enormous amounts of energy with it horrifying implications for weapons technology. The path to the world's first atomic bomb had been laid down. A few years later, the same science that had revealed the nature of the stars would be used to destroy hundreds of thousands of lives, and place the rest of us under an eternal threat of annihilation.

Chandrasekhar's Prediction

While all this was going on, a brilliant young Indian scientist was on a voyage from India to England to study with Eddington. Subramanyan Chandrasekhar was an exceptional student who had won a scholarship from the Indian government to study in England. On the voyage he had plenty of time to think. He was fluent in German and had read Einstein's papers on relativity theory, and

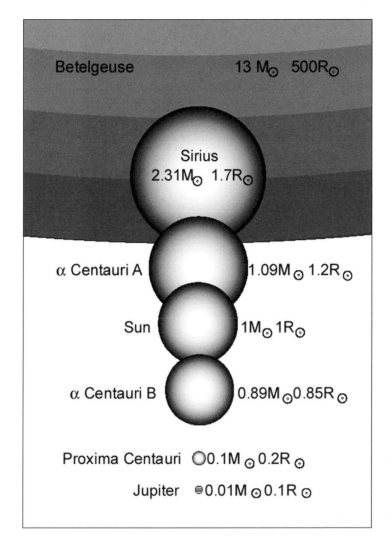

Figure 2a Comparison of sizes of typical stars. The Sun is described as a yellow dwarf, while Betelgeuse is a red giant. The nearest star to our Solar System is Proxima Centauri, a mere one tenth the diameter of the Sun. Jupiter – 11 times the diameter of Earth – is shown for comparison. Illustration by the author.

understood the intense conditions inside stars. Yet he was puzzled by the extreme conditions of stars that had used up their sustaining supply of nuclear fuel. What happened to them next? At the end of the life of our Sun, the nuclear fires that have kept it burning for over eight billion years will come to an end. Gravity will take over. The weakest of all four fundamental forces in the Universe is also relentlessly patient. Despite the titanic struggle nuclear fusion provided, the fight is now over: as the outer layers of the star drift off silently into space, gravity pulls what remains of the star's core inward. At some point the star stops contracting and becomes a white dwarf.

In Chandrasekhar's time, white dwarf stars were more than mere theory. Almost a hundred years earlier the Director of Konigsberg Observatory, Friedrich Wilhelm Bessel, had noticed the strange wobbling of the brightest star in the sky. He suggested that the star Sirius was wobbling from side to side as it orbited a common center of gravity with an invisible companion. Despite a lack of acceptance of his proposal, 18 years later the American lens maker Alvan George Clarke discovered the faint star predicted by Bessel. In 1914 American astronomer Walter Sydney Adams was able to photograph the spectrum of this peculiar star and found it to be just as hot as its companion Sirius. For a star to be this hot and yet so amazingly faint, there was only one explanation: it was tiny. Thanks to Rutherford's work on revealing the nature of the atom – recall that it is largely empty space, a tiny nucleus surrounded by a cloud of electrons – it was possible to comprehend matter being broken down and compressed in such a way that an entire sun could be compressed into something the size of planet Earth.

As the core of the star collapses at the end of its life, the electrons are stripped from the nuclei of atoms to form a kind of nucleus-electron soup called degenerate matter. However, no electron can occupy the same quantum space as any other electron, a phenomenon known as the Pauli exclusion principle. The core reaches a point where the electrons resist further collapse. It's a bit like trying to squeeze a bag of billiard balls: once they're all touching each other you can't make the bag any smaller. At a diameter approximating that of the Earth, the star's core halts its contraction and hovers between life and death. Although tremendously bright at the surface – at least when they're young – white dwarfs are incredibly faint in the sky owing to their small size. With no nuclear reactions to keep it burning, the initially hot white dwarf slowly radiates its remaining heat into space. Over billions of years it fades from white, through yellow, red and eventually black. As massive as a small star and as black as the night, any remaining planets will continue to circle the invisible stellar corpse in silent darkness. Forever.

For Chandrasekhar the question was not so much how could a star become so small, but rather what stops it becoming smaller and what happens at such densities? As Chandrasekhar thought more and more about the problem, he realized there is a limit to how massive a white dwarf can be. Note that we say massive, not big. It's a peculiar thing about white dwarfs that the more mass they have, the tinier they shrink. If we were able to add more mass to a white dwarf, its gravity would naturally increase pulling the stellar corpse ever tighter. Chandrasekhar's brilliance was to calculate, using nothing more than pen and paper, that beyond a mass of about 1.44 times the mass of the Sun,[8] a white dwarf undergoes further catastrophic collapse. The repelling influence of the electrons is overcome by gravity. What puzzled Chandrasekhar was what does the white dwarf collapse into?

By the time he arrived in England, he was ready to pursue this problem. But

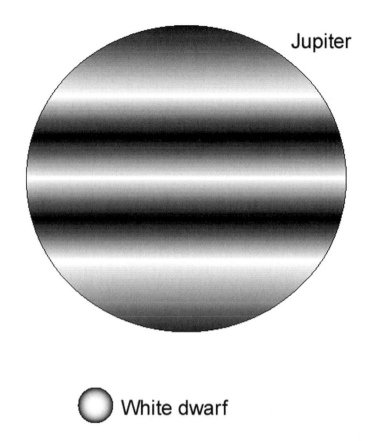

Jupiter

White dwarf

Figure 2b Size of a white dwarf star compared with Jupiter. Although they contain the mass of a star, at 1% the diameter of the Sun, a typical white dwarf is about the size of Earth. Could there be smaller, denser objects still waiting to be discovered? Illustration by the author.

the reception his theory received from Eddington and other British scientists was less than encouraging. Eddington himself would never accept Chandrasekhar's conclusions. Chandrasekhar presented his theory at a meeting of the Royal Astronomical Society during a lecture in January 1935. Following his lecture, Eddington proceeded to rubbish the whole idea. In one of the most famous arguments in the history of science, Eddington completely dismissed Chandrasekhar's findings and did so in a particularly harsh manner. 'I think there should be a law of nature to prevent a star from behaving in this absurd way' he declared to the audience. No one contradicted the great man. So revered was Eddington that his opinion held up research into the nature of small, dense stars for decades. As for Chandrasekhar, on the verge of predicting neutron stars, he felt Eddington's rebuke had ruined his chances of ever gaining tenure at a British university and moved to the United States where he joined the staff of the University of Chicago. He stayed there for the remaining 58 years of his scientific career and became one of the most important astronomers of the twentieth century.

Landau's Prediction

Speculation on the collapse of stars didn't end there. The same year, a young physicist, Lev Davidovitch Landau, had just moved back to his native Russia to take up two posts, one as head of the Theory Division of the Ukrainian Technical Institute in Kharkov and the other the chair of theoretical physics at the Kharkov Institute of Mechanical Engineering. Landau had been a child prodigy in mathematics, having completed his secondary schooling at the age of 13. He began studying mathematics, physics and chemistry at Baku University (later called Kirov Azerbaijan State University) the following year, and two years later moved to Leningrad State University from which he graduated in 1927. It was in Kharkov in 1934 that Landau received his doctorate in Physical and Mathematical Sciences. Landau's work covered a range of extreme physics, from atomic collisions to astrophysics, low-temperature physics to quantum field theory. One of his most important contributions was his explanation of the strange behavior of liquid helium at extremely low temperatures. At just a couple of degrees above absolute zero, helium becomes a superfluid, a bizarre form of matter which flows with no friction. The superfluid engages in some strange behaviors, such as climbing into or out of a beaker until the level of helium outside the beaker matches that inside the beaker. Landau devised a theory to explain the superfluidity of helium for which he won the 1962 Nobel Prize for Physics.

In later years, Landau would be imprisoned by the Stalinist authorities for being a German spy. Conditions in prison were such that, after a year, he doubted he would live more than another six months. Russian and foreign academics petitioned his release with the leading academic of the time, Peter Kapitza, and the German Niels Bohr, both writing to Stalin to ask for Landau's release. Kapitza in fact threatened to stop all research in protest at the imprisonment. Landau was released from prison in 1939 and given a position of research fellow at the Institute of Physical Problems.

Shortly before Chadwick unveiled the neutron, Landau was formulating his prediction of a star that had collapsed so far under its own gravity as to achieve an astonishing density. Independent of, and earlier than Chandrasekhar, Landau also showed that there is an upper limit for a white dwarf, being about 1.5 solar masses. Unaware of Chadwick's or Chandrasekhar's work, Landau conceived of a star that had achieved the density of an atomic nucleus and thus named it a 'nucleus star'. Landau worked out its properties, realized there must be a maximum mass and then drew the conclusion it could not exist since the idea of such a collapse flew in the face of stars with solar masses. While Landau's nucleus stars were not neutron stars in the modern sense, Landau was the first to predict the existence of stars of such density, and it is a little surprising that his prediction has gone so unacknowledged in the decades since it was made.[9]

Zwicky and Baade's Prediction

The first widely accepted prediction of ultra dense stars was that offered by two astronomers working in America: Fritz Zwicky and Walter Baade. Baade was one of the most important observational astronomers of the twentieth century. German by birth, he worked at Hamburg Observatory from 1919 to 1931, and then moved to the United States to work at Mount Wilson Observatory. In 1917 Mount Wilson became home to the world's largest astronomical instrument, the 2.5 meter Hooker telescope. During the war years Los Angeles and the San Gabriel Valley were subject to partial blackouts. Most other Mount Wilson observers were away working on weapons development, but Baade, being an enemy alien, was confined to Los Angeles. From April 1942 all enemy aliens were subject to a military curfew from 8pm to 6am, but the Director of the observatory, Walter Adams, appealed to the government and managed to get special permission for Baade to stay out after dark on the understanding that it was in the interests of professional work. So there he was: one of the world's most talented observers almost alone at the world's largest telescope under blacked out skies.

Baade put the time to good use and made his most important contribution. He turned the 2.5 meter Hooker telescope to the Andromeda galaxy, M31 and its companion galaxies M32 and NGC 205, and using recently acquired photographic plates, which were sensitive to the red end of the spectrum, Baade was able to resolve red giant stars. When Baade studied the developed plates, however, what were prominent by their absence were the red and blue supergiant stars which are plentiful in the spiral arms of the Milky Way. The brightest stars were yellow giants. What Baade had discovered was that there were two distinct groups, or populations of stars in the galaxy: Population I stars consist of young stars that are rich in metals,[10] including the bright, red and blue supergiants. They lie mainly in the spiral arms of the galaxy. Population II stars, in contrast, are found mainly towards the hub of the galaxy or out in the halo, and also make up most of the stars in elliptical galaxies. Population II stars are old and were later shown to be metal poor, the first generation of stars in a galaxy. The implications of Baade's discovery were profound and set the stage for studies of galaxy and stellar evolution for years to come.

Zwicky,[11] too, was an amazing scientist who made many important contributions, not only to astronomy but also a wide range of other disciplines. Born in Bulgaria to Swiss parents, Zwicky spent most of his working life in the United States. After receiving a PhD in physics in 1922 from the Swiss Federal Institute of Technology, Zurich, he moved to the United States where he served on the faculty of the California Institute of Technology, Pasadena, from 1925 until 1972. Zwicky's contributions ranged from the prediction of how dark matter in the Universe can be mapped using a phenomenon known as 'gravitational lensing' to developing some of the first jet engines. Not many scientists are intellectually decades ahead of their time, but certainly Fritz Zwicky was a remarkable example of such a person. At the personal level, Zwicky was a loud character with a very strong accent, despite almost five decades of living in the United States. He was bombastic and rather self-opinionated, and had a low opinion of many of his colleagues. One of his favorite insults was to refer to people he didn't approve of as 'spherical bastards' because, he explained, they were bastards no matter which way you looked at them.

Above all, Zwicky was an observational astronomer: although he did some important theoretical work, he is best known for his observational contributions. For example, he was one of the first to show that the apparent concentrations of galaxies (first pointed out by Herschel) were in fact true clusters. Using the 48 inch Schmidt telescope on Mount Palomar Observatory he discovered many galaxy clusters. Between 1937 and 1941 Zwicky discovered 18 supernovae in other galaxies at a time when the sum total of supernova discoveries by all other astronomers was 12.

In 1934 Zwicky and Baade collaborated on one of the most important predictions in the history of astronomy. They had witnessed a number of major events in physics: the development of quantum theory, the discovery of the neutron, and the growing knowledge of stellar evolution. But a major unresolved puzzle at the time was the nature of cosmic rays, charged particles that arrive on Earth from space. (To this day the origin of cosmic rays remains ambiguous.) Rejecting the accepted suggestions of the day, they offered an alternative. The energy and random distribution of cosmic rays seemed to indicate a more sporadic source. What if, they said, cosmic rays were the result of supernovae, of the death throes of massive stars undergoing gravitational collapse? The intensity of cosmic rays can be explained by matter collapsing to ever denser states, releasing energy as it does so. Such a titanic explosion was surely energetic enough to produce the observed cosmic rays.

Baade and Zwicky went further still. One of the by-products of such a supernova would be the creation of a bizarre new type of star.

> With all reserve we advance the view that a supernova represents the transition of an ordinary star into a neutron star, consisting mainly of neutrons. Such a star may possess a very small radius and an extremely high density. As neutrons can be packed much more closely than ordinary nuclei and electrons, the "gravitational packing" energy in a cold neutron star may become very large, and, under certain circumstances, may far exceed the ordinary nuclear packing fractions. A neutron star would therefore represent the most stable configuration of matter as such.[12]

That 'very small radius' was a few tens of kilometers, a hundred times smaller than a white dwarf. Chadwick's 1932 discovery of the neutron had led inexorably to the conclusion that the nuclear forces at work in stars championed by Eddington were only the beginning. Out of stellar death came a new incarnation for massive stars. The neutron, freed from the electrical repulsion of electrons and protons that resisted collapse in white dwarfs, allowed dense stellar cores to collapse still further under the influence of gravity. In 1939 the details of the structure of neutron stars was published in the journal *Physical Review*. The authors were the American J. Robert Oppenheimer and his student George M. Volkoff. The former was destined to become the leader of the Manhatten Project that developed the world's first atomic bomb using the penetrating properties of the neutron.

As intriguing as neutron stars were, the problem remained as to how anyone could possibly observe such a thing. Detecting a white dwarf was one thing, but neutron stars were hundreds of times smaller, and correspondingly fainter. The discovery of the first white dwarf had pushed optical technology to the limit. No one could imagine a way of ever detecting such a faint object as a neutron star. As a result, the neutron star was destined to remain a theoretical curiosity for more than thirty years. But as so often happens in science, their detection lay not in the development of theory, but rather in the unpredictable advances of technology, which opened new windows onto the Universe.

References

1. Daughter of Pierre and Marie Curie.
2. Helium nuclei, now known to consist of two protons and two neutrons.
3. γ rays are even more energetic and penetrating than the more familiar X-rays used by the medical profession to explore the nature of broken bones.
4. Quoted in Pais (1986, p. 398).
5. The word 'neutron' had been used as early as 1898 to describe various particles, including proton-electron pairs.
6. Quoted in Pais p. 398.
7. Named after a Russian physicist Piotr (Peter) Leonidovich Kapitza.
8. It depends on the composition and rotation of the white dwarf, but the details don't matter to our story.
9. Thanks to Kenneth Brecher for his insights into this piece of history.
10. Astronomers refer to anything heavier than lithium as 'metals': they are the product of highly evolved stars and supernovae.
11. Much of the biographical data on Zwicky was published in *In Search of Dark Matter* by Ken Freeman and Geoff McNamara, Springer, 2005.
12. Baade, W. & Zwicky, F. 1934, 'Cosmic Rays from Super-Novae' *Proc. Natl. Acad. Sci, USA*, Vol. 20, p. 259.

3   'A New Window'

'Twinkle, twinkle quasi-Star
Biggest puzzle from afar
How unlike the other ones
Brighter than a billion suns.
Twinkle, twinkle quasi-Star
How I wonder what you Are.'
George Gamow, *Quasar* 1964.

The Rise of Radio Astronomy

The night sky is a window onto the Universe. During the day the sunlight is scattered by the air molecules and obscures our view. For astronomers, it's like trying to watch a film at the cinema with the lights on. But when the Sun sets, the sky darkens and we see Earth's celestial surroundings. Since prehistory, humans regarded the night sky as a hemispherical surface to which the stars were attached, but now we know better: with insight the dark dome of the sky disappears and we are free to peer deeply into space. The difference in the apparent brightnesses of the stars is a combination of intrinsic luminosity and distance, but generally speaking the fainter stars are further away. The further we look the more numerous the stars. At some point they cease to be individuals and merge to form the ghostly band of the Milky Way just as sand grains merge to form the smooth texture of a beach. Diffuse patches just visible to the naked eye are galaxies themselves far beyond our own. As technology advances, so does our perception, but there is no obvious end to what we will discover. The sky has neither surface nor limit: when we look into the night sky we are humbly yet irresistibly peering into the cosmos. Yet there is more to the Universe than meets the eye: there exists a universe invisible to our natural senses, and like so many major scientific developments, this new universe was unveiled completely by accident. Ironically, it was initially explored not by astronomers who had devoted their lives to watching the skies, but by radio engineers intent on winning a war.

Karl Jansky

As we saw in the last chapter, 1932 was a busy year in science: as Chadwick identified the neutron, Landau speculated on the 'nucleus star'. In that same decade, radio was becoming a popular source of communication and entertainment. But it wasn't without its problems. Bell Telephone Laboratories wanted to exploit radio transmissions for trans-Atlantic telephony, but a persistent hiss plagued what was then regarded as short wavelengths (10–20 meters). The task of finding out where this noise was coming from was given to a young radio engineer, Karl Guthe Jansky, who had joined the company in 1928. Although at the time not seen as a terribly spectacular occupation, the hiss he was investigating was our first glimpse through a new window onto the Universe. The new window would reveal not only the structure of the Milky Way, but a host of new astronomical phenomena including pulsars.

Jansky built an antenna that could detect radio waves with a wavelength of 14.5 meters. To allow him to determine the direction the static was coming from he mounted the antenna on a large turntable, the appearance of which earned it the title 'Jansky's merry-go-round'. After months of listening, Jansky could distinguish static from nearby and distant thunderstorms. But what intrigued him was a faint but steady hiss that seemed to come from nowhere in particular. Because the intensity of the static rose and fell each day Jansky initially thought he was receiving static from the Sun. Since Earth rotates once every 24 hours the Sun is carried across the sky at the same rate all year round. But after months of monitoring, it was clear that the brightest point was moving, not with a regular 24 hour cycle as would be the case with the Sun, but with the shorter period of 23 hours 56 minutes. The loss of four minutes a day is what the distant stars experience due to the Earth's orbit around the Sun. The difference was small, but it was enough to confirm that, not only was the source celestial, it was beyond the Solar System. The static that Jansky had discovered was coming from deep space.

Initially Bell Labs capitalized on the discovery and gained wide publicity. When Jansky proposed construction of a 30 meter diameter dish antenna for further investigation of the phenomenon, however, he received no support from his employer. As interesting as space-based static was for Jansky, as far as Bell Labs was concerned it posed no problem for their enterprises. They had their answer, and that was that. Jansky was given another project to work on, and he never again returned to radio astronomy.

Grote Reber

The radio sky was ignored for years. Then in 1934 a young Grote Reber read one of Jansky's papers and thought he knew the answer to making a more detailed investigation of the phenomenon. By constructing a parabolic dish it should be possible to study these strange radio emissions at higher angular resolutions, thus pinpointing where in the sky the emissions were coming from. Reber wrote to Jansky with his proposal, but for reasons that are unclear Jansky was not encouraging. Neither were the astronomers Reber contacted. As he put it during a 1992 interview:[1] 'The optical astronomers knew nothing about a radio set and so thought Jansky was wrong, while the electrical engineers believed Jansky's work but could see no point in it.' But not everyone thought it was a waste of time, and within a few years a Cambridge physics professor Gennady Potapenko and his student Donald Follard had confirmed Jansky's findings using a loop antenna, and later using a single wire antenna in the Mohave Desert. Despite Jansky's discouraging reaction, in 1936 Reber constructed a 31.5 foot fully steerable parabolic dish in the backyard of his parents' home. With the help of a local blacksmith, it cost a grand total of $1,300 and allowed him to study the radio sky at wavelengths of 74.9cm (~400 MHz) and 199.9cm (150 MHz). With this instrument, Reber became the world's first and only radio astronomer. Sadly, academics refused to accept his work because he had not done so under suitable university academic supervision.

Discrete Radio Sources

Among the most important developments were the discovery of sunspot groups at radio wavelengths, and the prediction and subsequent discovery of the all-important 21 cm line emission due to hydrogen. The latter was an incredibly important development in revealing the shape and size of the Galaxy we live in. Many popular books will tell you that looking in the direction of Sagittarius you are peering at the center of the Milky Way galaxy. In truth, this is like saying if I stare at the ground beneath my feet I am looking at Greenland (you can make adjustments for your local geography). Between us and the rest of the Galaxy lie vast clouds of interstellar gas and dust, and trying to map out our surroundings is like trying to navigate your way through a dense fog at night wearing sunglasses. But in the mid-1940s, the great Dutch astronomer Jan Oort wondered about the possibility of studying interstellar hydrogen (which constitutes the bulk of the stuff the Galaxy is made of) at radio wavelengths. He set his student Hendrik van de Hulst the task of investigating the possibility, and in 1945 van de Hulst came up with the answer. At a wavelength of 21 cm, he said, hydrogen emits a specific spectral line. If a radio receiver could be built to 'see' this wavelength alone, then the hydrogen would become visible, and the obscuring dust and gas of the Milky Way would become transparent. Van de Hulst in fact approached Reber with the proposal. Reber began constructing an antenna for the job but shortly afterward was offered a job with the National Bureau of Standards in Washington. He abandoned the endeavor. It would be another six years before the 21 cm line was finally discovered by Harold ('Doc') Ewen and Edward Purcell. With the veil of the interstellar gas and dust finally lifted, astronomers were able to map out the structure of the Galaxy revealing its spiral structure. For the first time in history, astronomers were able to peer beyond the confines of the local stellar neighborhood and explore the Galaxy.

Of greater importance to this story, however, was the gradual refinement in radio technology that allowed the discovery of *discrete* radio sources, which are sources of radio waves that are small and distinguishable from the background

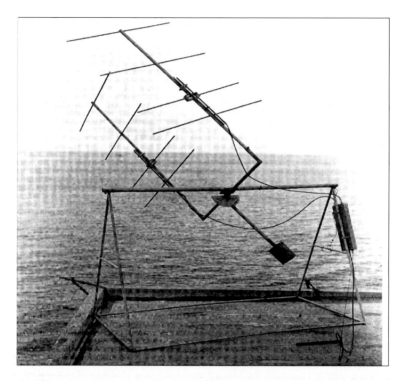

Figure 3a The original antenna at Dover Heights, east of Sydney. The 100 MHz twin-Yagi antenna positioned on a blockhouse at Dover heights and used for the finding of the first discrete sources using sea interferometry. In the orientation shown, the Yagis are being used for measuring the polarization of solar emission. For the work on the discrete sources, they were used in a parallel configuration and swung down to be parallel to the horizon. (Courtesy: ATNF Archives.)

radio glow. During the Second World War, radio technology went through a major evolution, and by the end of the war, radio engineers in England, Europe, the United States and Australia were turning their thoughts and antennas to the sky. The first intense radio noise from the Sun had been accidentally discovered by British coastal radar stations during the war and attributed to attempts by the Germans to jam their effective operation; it was James Stanley Hey who investigated the phenomenon and came to the correct conclusion that the Sun was emitting radio waves. It was also Hey (1946) who made one of the first meter-wavelength sky surveys, during which he discovered a strong discrete radio source in the constellation of Cygnus; he was, however, unable to locate its position with enough accuracy to attempt an optical identification.

Most of the initial important discoveries were made in Australia. Observing from Dover Heights on the coast east of Sydney, John Bolton, Gordon Stanley, and Bruce Slee used equipment set up on the cliffs overlooking the Tasman Sea. Just as the rising Sun glistens on the water, as each radio object rose over the

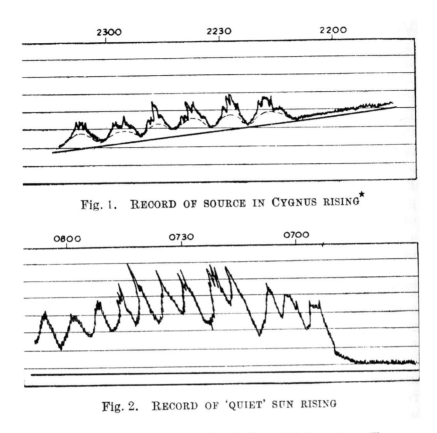

Fig. 1. RECORD OF SOURCE IN CYGNUS RISING*

Fig. 2. RECORD OF 'QUIET' SUN RISING

Figure 3b Interference fringes produced by the Dover Heights antenna. The upper trace shows the sea-interferometer fringes from the finding chart recording of Cygnus A. The fluctuations on the tops of the fringes are the well-known ionospheric scintillations. The lower trace is a comparison chart recording from the rising Sun, using the same antenna in Figure 1. Note that the minima of the Cygnus A fringes fall almost to zero, while the solar fringes do not have deep minima, indicating that the angular size of the quiet Sun is much greater than that of the source Cygnus A. (Courtesy: ATNF Archives.)

ocean's horizon its image was reflected off the ocean, creating two sources of radio waves that alternately reinforced and canceled each other.[2] The effect can be demonstrated by dropping two stones in a still pond and watching the two sets of waves alternately cancel out and reinforce as they lap across each other. Radio waves behave in much the same way. By adjusting the radio antenna so that it received both sets of waves, the engineers were able to approximately locate the celestial position of the source from the strength, rising time and periodicity of the resulting interference pattern. This technique, known as interferometry, was to play an ever more important role in radio astronomy in the decades to come.

The group at Dover Heights made it their first task to re-detect the source in

Cygnus first announced by Hey in 1946, and to find a much more accurate position for it. They detected its sea-interference pattern in June 1947 and named the source Cygnus A, but there was nothing at its position save a couple of faint stars. But where there was one distinct radio source, Bolton and his team reasoned, there must be others, and they began to survey the skies with their sea-interferometer. They soon found another one in Taurus on November 6th 1947 and by March 1948 had found a further six sources scattered across the sky. Radio sources are not star-like in appearance, however: while it is easy to point to an individual star, the location of radio sources was at first rather vague. Like Cygnus A, none of the other radio sources matched objects seen at visible wavelengths. The strongest of the sources, Cygnus A, was not identified until the early 1950s by a team led by Francis Graham-Smith in Cambridge, whose more accurate position coincided with a rather faint nebulous region. Walter Baade and Rudolf Minkowski used this position as the basis of a spectroscopic analysis of the visible object, identifying it as a galaxy and concluding that it emitted most of its energy in radio wavelengths.

To improve resolution of radio sources, Bolton and Stanley traveled to New Zealand in mid-1948. They had two goals. One was simply being able to see the sources set over a western ocean horizon allowing them to more accurately determine the position of the sources. The second is that New Zealand had higher cliffs than anything on the east coast of Australia and it was hoped that the greater elevation above the sea would increase the angular resolution, that is produce a sharper radio 'image'. Carting an aerial around on a large trailer they selected two cliff-top sites with elevations of around 300 meters (as opposed to Dover Heights at 79 meters). Despite frequent shocking weather conditions, four months of observation and analysis revealed the positions of four of the new sources to within a few minutes of arc (about a sixth the diameter of the full Moon). Two of the sources coincided with optical images of galaxies: the position of Virgo A matched a giant elliptical galaxy called M87, the second brightest member of the 2,500-strong Virgo cluster of galaxies. Centaurus A was identified with NGC 5128 which later proved to be a strong source of X-rays, infrared and gamma radiation, and the nearest of the so-called active galaxies.[3] A third radio source, Taurus A, matched a thousand year old supernova remnant called the Crab Nebula.

The Radio Crab

The Crab Nebula is so named because the ragged outline very loosely resembles a crab. The tiny patch of light, just visible in small telescopes, coincides with the location of a 'new star' seen by Chinese and Japanese astronomers in 1054, who recorded its visibility even during daylight hours for more than three weeks. By the time Bolton made his discovery the various pieces of the stellar evolution jigsaw described in the first two chapters were starting to fall into place. The 'new star' was in fact a supernova, the violent and spectacular death throes of an old, high mass star. Almost 900 years later it was still beaming radio waves across the Galaxy. The Crab was the first visual object (other than the Sun) to be identified as a radio source, and hence John Bolton, Bruce Slee and Gordon Stanley are now recognized as the first to identify radio sources with optical sources. It also marked the coming of age of radio astronomy: at last it was shown that radio astronomy was able to make discoveries completely independent of optical astronomy. A new era in scientific research had begun.

Australians weren't the only players in the new field of radio astronomy. In Cambridge, England a leading expert on upper atmospheric physics, John Ratcliffe, had returned to Cambridge University at the end of the war. Ratcliffe spent the war years on radar work, but when he returned to Cambridge he recruited a number of young radio scientists including Martin Ryle, Francis Graham-Smith[4] and Antony Hewish. For a year the men salvaged as much British and German radar equipment as they could and eventually began radio observations. After initially observing radio emissions from sunspots, the team turned their attention to discrete radio sources. While the Australians were the first to discover these sources and make the initial sky surveys, the British team carried out the first large scale surveys of the sky in search for new objects. The first Cambridge Survey (there are now nine published Cambridge surveys) made use of a technique developed by Ryle called 'phase switching'. This involved taking the signals from two widely spaced aerials and delaying one of them so that it is alternately in and out of phase with the signal from the other. The result of this technique is easier identification of the point radio sources against the background. Using this technique the Cambridge team were able to catalog 50 sources. Four years later, in 1952, Ryle led a team that made use of a 'double interferometer' consisting of four cylindrical aerials placed in the corners of a large field a few kilometers west of Cambridge. After three years the astronomers had cataloged 1,936 radio sources in the 2C catalog. Yet it was the third Cambridge catalog, or 3C, that was to reveal an entirely new astronomical phenomenon.

Quasars

There was a type of radio source that stood out against all the others. Seen against the sky they were tiny yet bright, the radio equivalent of stars and so astronomers dubbed them quasi-stellar radio sources, then simply quasars. Now a radio telescope's 'vision' is decidedly blurry compared with optical telescopes and for a while it was difficult to know not only how point-like quasars were, it wasn't even clear precisely *where* they were. Both aspects were clarified in 1962 when the Moon passed in front of the brightest quasar, 3C273.[5] As the Moon orbits the Earth it passes in front of any celestial object more distant than itself, an event called a lunar occultation. Because the Moon is an airless world, seen from Earth, its edge, or 'limb', is sharply defined. Whenever the Moon occults a point object like a star, the star simply winks out rather than fading slowly, as is the case with objects with a discernable diameter like planets in the Solar System. Originally the technique of lunar occultations was used to plot the position of the Moon: knowing accurately the positions of the stars the motion of the Moon could be accurately plotted. Knowing the precise location of the Moon means you can also do the reverse: by recording the time that a radio source disappeared behind the Moon, it was possible to unambiguously plot the location of the radio source so it could be matched with visible objects. In the early 1960s a few attempts had been made to pin down these strange radio sources with limited success. The real breakthrough came in 1962 using the 64 meter Parkes radio telescope in western New South Wales.

Parkes and 3C273

It just so happens that once every twenty years a series of occultations of 3C273 becomes visible from the southern hemisphere. Cyril Hazard, then working with the new astronomy group at Sydney University, booked time on the newly completed Parkes telescope to observe the events. The first observation was inconclusive. There was another occultation of 3C273 yet to be observed, but there was a problem. The huge diameter of the Parkes dish compared with its relatively squat support tower meant that the dish could not be tilted more than 60 degrees from the zenith. As the enormous dish reaches an elevation of 30 degrees above the horizon and the rim of the dish approaches the ground, an alarm sounds and the telescope automatically stops. The trouble was that at the time that 3C273 was to emerge from behind the Moon, the pair of objects would be a little lower in the sky than 30 degrees. This was to be an important observation, not only for the radio astronomers but for the Parkes telescope itself. It was hoped that the newly completed telescope would begin its career with some major discovery, and quasars held the promise of something big. Bolton and his colleagues weren't going to let such a minor restriction as a safety protocol get in the way of a potentially important astronomical discovery. Not only did they remove the telescope's safety devices, but Bolton also took a grinder to the telescope, cutting away some of the housings of the bearings allowing the dish to tilt another degree closer to the horizon. In readiness for the observation all the roads leading to the observatory were closed, electrical equipment not essential turned off, and electronic equipment needed for the observation itself duplicated. In the early evening of August 5th 1962, 3C273 was occulted by the Moon. At the time of reappearance the Parkes telescope's rim was almost touching the ground, but they saw it all. So important were the observations that Hazard and Bolton returned to Sydney carrying duplicate copies of the records of the event on separate planes, just to be sure.

Speed and Distance

When the radio observations were analyzed, it was clear that 3C273 was not one point but two. Subsequent comparison with photographs of the region taken with the 200 inch Hale telescope showed that the radio source coincided with an unremarkable blue 'star' with a faint jet extending from it. At this stage, most astronomers believed that these 'radio stars' were just that: peculiar stars that inhabited the Milky Way just like all the other stars. The next step was clear: find out more about this 'star' that looked so peculiar in photographs and had such strong radio emissions. As we saw in Chapter 1, the way to analyze stars is through their spectra. To learn more about the quasars, a Dutch astronomer Maarten Schmidt working at Palomar observatory in the United States used the 200″ telescope, the largest in the world at the time, to successfully obtain the spectrum of 3C273. It was expected that an analysis of the lines in the spectrum would tell astronomers about the star's composition, motion, and so on. But when Schmidt looked at the arrangement of the lines, nothing made sense, at least at first. Comparing the arrangements of the lines with those in tables created from observations of other stars revealed nothing. Could the star contain previously unknown elements? What was going on? It took Schmidt four months to discover the solution. But to understand Schmidt's insight we need to take a minor astronomical detour.

It had been known for decades that the arrangement of spectral lines in the spectrum of a star or galaxy revealed not only its composition, but also its velocity. Light behaves in a similar way to sound when it comes to motion. You may have had the experience of hearing a vehicle with a siren such as an ambulance at first approaching you and then receding into the distance. When the siren is coming towards you it sounds higher pitched than when it is moving away. This is called the Doppler Effect and results from the relative motion of the source of the sound waves and the listener's ear. As the ambulance is moving towards you, the sound waves are compressed resulting in a higher pitch; when the ambulance is moving away from you the sound waves are stretched out, giving them a lower pitch. Like sound, light is made of waves, so if a star is moving towards us, its spectral lines are shifted towards the shorter wavelength, blue end of the spectrum, the equivalent of a higher pitch. When a star is traveling away from us its lines appear shifted towards the red end of the spectrum. This is how astronomers are able to plot not only the position but also the motion of stars in the Galaxy.

Now when it comes to galaxies well beyond the Milky Way's neighborhood a phenomenon with a similar appearance emerges. One of the pioneers of studying the spectra of galaxies was Edwin Hubble. Hubble observed thousands of galaxies, patiently recording their spectra and then analysing them. His observations revealed two important astronomical phenomena. Firstly, almost all the galaxies beyond the Milky Way showed a redshift in their spectra. This meant that they were receding, that is moving away from us.[6] It is important not to be lulled into thinking that because all the galaxies are moving away from us, the Milky Way is at the center of the Universe. On the contrary, it confirms that we are in no special location in the Universe and that each galaxy is receding from every other galaxy. This mutual recession of the galaxies is often illustrated by imagining sultanas in a fruit cake cooking in an oven: as the cake rises, each sultana moves away from the others, carried by the expanding cake. The second thing Hubble demonstrated was a simple law that relates distance to recession velocity. An important implication of the mutual recession of galaxies is this: the faster a galaxy is receding from you, the further away it is. For example, imagine three galaxies equally separated along a line and all moving away from each other at the same rate. Now imagine us living in a galaxy at one end of the line. From our perspective, the middle galaxy will be moving away from us just as we would appear to be moving away from inhabitants of the middle galaxy. But if we were to measure the recession of the second galaxy (which is twice as far away) it would appear to be moving away twice as fast as the nearer galaxy. This is because it is receding from the nearer galaxy, which is itself receding from us. In other words, the greater the recession velocity of a galaxy, the further away it is. Because the recession velocity is proportional to the distance, the ratio of the velocity to the distance is approximately constant as we go from galaxy to galaxy. This constant is called the Hubble constant, which has also been notoriously difficult to determine accurately. For one thing, it is very difficult to measure precise distances to galaxies. Nonetheless, no one could argue the fact that the galaxies were a long way away.

Hubble paid little attention to the theory behind just why the Universe might be expanding; however Albert Einstein was very interested in the idea. General relativity envisioned space not as a blank chessboard on which the various pieces moved, but rather a tangible substance to which everything was attached. Like galaxies painted on a huge rubber sheet, the bending, flexing and stretching of space influenced the position and motion of everything within it. The recession of the galaxies was not the motion of these objects through space; it was evidence of them being carried along by the expansion of space. This in itself was an important prediction of general relativity that Einstein at first could not accept. He was so concerned about the idea of an expanding or contracting Universe predicted by his theory that he introduced the famed cosmological constant to keep his universe still. It was only when Hubble announced his discovery of the recession of the galaxies that Einstein realized his prediction without the cosmological constant had been right all along.

Of course in order for an object to feel the effects of the expansion of the Universe they have to be at a very great distance. That's why relatively nearby objects like the stars of our Milky Way galaxy, or even the 2 million light year distant Andromeda galaxy, are all relatively unperturbed by the expansion of space. It's only over cosmological distances that a high redshift indicates a large recession velocity and hence a great – that is cosmological – distance. Until Schmidt looked at the spectrum of 3C273, the furthest objects known were the nearby galaxies, vast 'island universes' that shone with the collective brilliance of their billions of stars. Astronomers were still coming to terms with the immensity of the Universe revealed by Hubble's law, and to this day make maps of the Universe by measuring the redshift of individual galaxies, converting this redshift to distance, and plotting the position of the galaxies in space.

Schmidt's Insight into Quasars

Now we can return to Schmidt and his remarkable discovery. In the 1960s measuring the redshift and distance of a galaxy was pretty well understood, and yet that's why it took Schmidt so long to understand the spectrum of the quasar 3C273. After months of puzzling over the forest of spectral lines, he realized that the lines he saw were those produced by none other than hydrogen, the simplest and most abundant element in the Universe. What had perplexed him was their position in the spectrum: they were further towards the red end of the spectrum than anything anyone had seen before. The extreme redshift of the lines could only be produced by one thing: immense velocity away from the Milky Way, and that meant the object was at an immense distance. When Schmidt measured the redshift he found it to be at an unprecedented $z = 0.158$ of the rest wavelength, implying a distance of a staggering 2 billion light-years.[7] But to be detectable at either visible or radio wavelengths from such a distance meant that 3C273 had to be unbelievably bright, and that meant a supply of energy on a scale not previously known. Clearly quasars were a new type of astronomical phenomenon that cried out for astronomers' attention.

The Hunt for Quasars

By the end of 1964 there was a total of forty quasars known, some with redshifts of 200% implying distances of 10 billion light-years, effectively doubling the size of the known Universe. The implication was that this was the tip of the iceberg. Such an important astronomical discovery beckoned further investigation, and if astronomers couldn't explain what they were, at least they could go looking for more. We have already seen how the Cambridge group had developed expertise in creating radio telescopes for surveys of the radio sky. The Cambridge surveys were revealing ever more detail. Quasars began to feature prominently in the searches, but identifying quasars was difficult. After all, there is a limit to how much of the sky the Moon passes through, limiting that particular identification technique. One of the astronomers to pioneer the use of radio interferometry at Cambridge, Antony Hewish, had an idea for detecting new quasars seen anywhere in the sky; however, it would require a new technique, using a new design of radio telescope. And he would need assistants to build and run his new design. One of the staff he hired was a young PhD student, Miss S. Jocelyn Bell.

References

1. Raymond Haynes and Roslynn Haynes, quoted in Haynes, *et al.* 1996.
2. Bruce Slee has offered to supply a picture of the original antenna system and the resulting interference fringes.
3. A galaxy that has unusually strong energy emissions.
4. Graham-Smith would later become England's Astronomer Royal, preceded in the role by Ryle.
5. 3C273 indicates it was the 273rd source in the 3rd Cambridge catalog.
6. Although Hubble identified this universal expansion, we should note that his observations of receding galaxies were preceded by Slipher, who was the leading authority of galactic velocities of the time.
7. To put this in perspective, the Milky Way is approximately 130,000 light-years across; the nearest large galaxy, Andromeda is 2 million light-years distant.

4 'Scruff'

Jocelyn Bell cycled out to the observatory as she had done many times before. Looking more like a vineyard than a radio telescope, the Interplanetary Scintillation Array – Hewish's brain child to search for quasars – collected radio signals from space and recorded them as a series of squiggly lines on a paper chart. Among Bell's jobs were feeding the recorder with paper and filling the ink wells for the recording pens. Her most important job, however, was to sift through the recordings looking for signs of quasars. Computers were around at the time, but it was thought a trained human was better at distinguishing between interference and quasar signals, at least until the astronomers became familiar with the behavior of the telescope. After analysing a hundred meters or so of the chart recordings, the 24-year-old Ph.D. student became rather good at her task. It was impractical to monitor and analyze the recordings in real time, and so Bell waded through the kilometers of paper chart recording after it had accumulated. Then one day in August 1967, she noticed what she would later describe as 'scruff' on the recording. Unlike any natural signal she had seen before, it didn't look artificial either. Puzzling. Although it was probably nothing, she marked it with a question mark and went back to searching the recordings for quasars.

The Search for Quasars

Quasars were a new and celebrated celestial phenomenon. These bright radio beacons were as pin point as the stars and yet more distant than the galaxies. What could be so distant yet shine so brightly? What could generate so much power that it beamed across the immense gulf of the Universe? Answers to these and a hundred other questions kept astronomy buzzing, but the next logical step was to see just how many quasars were out there and what they were like. Fortunately, the star-like appearance of quasars led to an ingenious method of finding more of them. If you've ever looked at the night sky and known which bright points were stars and which very few were planets, you may have noticed that the planets don't twinkle. The simple reason is that, seen from Earth, the planets have a small but definite diameter. In contrast, despite the fact that the stars are so very much larger than planets, they are so far away they appear as points of light. Now when a beam of light from a point source comes streaming through the Earth's atmosphere, it is refracted, or bent, in unpredictable ways. This scatters the light around a tiny bit causing the star to twinkle in the sky. In the mid-1960s it had been noticed that the images from point radio sources beyond the Solar System also twinkle, but for a different reason. Earth's atmosphere only extends a tiny distance into space, but throughout the Solar System is an ever-expanding haze of charged particles streaming away from the Sun called the solar wind. Imagine this stream of particles not so much as a homogenous stream, but rather as huge irregular blobs of tenuous gas belching away through the Solar System creating turbulent streams in all directions. Starlight is unaffected by the solar wind, but radio waves are a different story. As radio waves pass through the solar wind they diffract, a different kind of bending from refraction but the effect is similar: it causes a kind of twinkling of the radio sources called interplanetary scintillation. Compact radio sources scintillate more than extended ones, and lunar occultation observations had shown quasars were remarkably compact radio sources. Antony Hewish reasoned that here was a way to find more quasars: search the sky for radio sources – itself a worthy goal – pick out the ones that twinkle, and you've found yourself more quasars. But to do this he would need a radio telescope of unprecedented sensitivity and resolution. Hewish knew how to achieve both.

Antony Hewish and the IPS Array

Antony Hewish had been involved in radio work during the war years, concentrating on devices to counter airborne radar at the Royal Aircraft Establishment and also at the Telecommunications Research Establishment where he first worked with Martin Ryle. Ryle himself was a radio astronomy pioneer, one of the first to carry out interferometric radio studies and was the driving force behind a group at Cambridge. Originally known as the Radio Astronomy Group (now the Cavendish Astrophysics Group), they were the creators of the Cambridge catalogs of radio sources and the discoverers of interplanetary scintillation. Hewish was motivated to study radio astronomy not only by his experiences with electronics and antennas during the war years, but also by an inspiring course in electromagnetic theory he completed as an undergraduate. Suitably motivated, Hewish joined Ryle immediately after completing his undergraduate degree at Cambridge (it had been interrupted by the war) and completed his Ph.D. in 1952. Hewish wasted no time in making major contributions to the fledgling field of radio astronomy. In particular, he was interested in how radio waves behaved as they passed through transparent though inhomogeneous media, including the solar wind. Hewish showed how interplanetary scintillation could be exploited to produce high resolution maps of the radio sky. As we saw in Chapter 3, larger diameter telescopes are needed in order to see fine details in the sky, but bigger means more money and work.

Figure 4a The Interplanetary Scintillation (IPS) Array used by Hewish and Bell-Burnell to discover pulsars. The array consists of 1,000 posts hammered into the ground, and strung with almost 200 km of wire, making a total of 2,048 aerials. (Courtesy Graham Woan.)

Hewish got around the problem by designing a phased-array antenna that not only simulated the resolution of a radio telescope 1,000 kilometers across using a much smaller area, but one that would daily survey the entire visible sky.

The telescope Hewish had in mind would use a new technique for detecting compact radio sources by being able to detect rapid fluctuations in radio signals so that anything twinkling at radio wavelengths would stand out from the 'steady' appearance of extended sources. This was how the quasars would be detected. He obtained funding for the telescope in 1965 and it took two years to complete. A team of five built the array, including Jocelyn Bell. Now this radio telescope isn't like the gleaming white dishes we see today. Starting out with over 1,000 posts standing upright in the ground, sledge hammered in by several vacation students during a summer break, Bell and her colleagues then strung almost 200 kilometers of wire and cable among the posts to make up a total of 2,048 aerials. The entire array covered an area of just under two hectares, the equivalent of 57 tennis courts. (Such an intricate arrangement of obstacles is any lawn mower's nightmare, and so sheep are employed to keep the grass around the array to a manageable level!) It's pretty clear that a construction like this isn't able to point at different parts of the sky, but that's not a problem since the Earth does the pointing for you. As the Earth rotates it carries the beam (the stream of radio waves being received by the telescope) across the sky while clever electronic techniques allow the astronomers to 'look' from directly overhead down to within about 30 degrees of the horizon.

July–August 1967

The IPS Array was still several months from completion when Bell began using it to search for scintillating sources. She had sole responsibility for operating the telescope and analysing the data under Hewish's supervision. As the radio sky moved overhead, the telescope scanned four different beams and recorded their results on four, three-track pen recorders. Hewish's instructions were clear and precise: scrutinize the recordings and record every scintillating source. 'I instructed Jocelyn to keep a quick-look sky chart plotting the position of each scintillating source whenever it appeared on the weekly records,' he explained to me. 'Genuine sources would repeat at the same positions while interference would occur randomly. My technique for measuring angular sizes needed observations at different solar elongations, hence the requirement for repeated observations.' Bell first encountered the 'scruff' about six weeks into the program. Some time later she saw more scruff on the recording and knew she had seen this type of signal before. Checking through earlier records she found the original scruff and plotted its celestial position at Right Ascension[1] 19 hours 19 minutes. Many would have dismissed the tiny patch of scribble as interference, but to Bell it didn't look artificial. At the same time she had become familiar with scintillating sources and this didn't look like one of those either. Besides, it had appeared near midnight in a patch of sky least affected by the solar wind; if it was a quasar, it shouldn't have been twinkling. Suspecting this was a point source, Hewish decided to have a closer look. The way to do this with a radio telescope was to simply move the paper faster under the pens, thus spreading out the signal over a longer strip, the radio equivalent of making a photographic enlargement. Hewish instructed Bell to install a fast recorder so they could examine the source in more detail the next time it appeared. 'I initially thought the scruffy signal might have come from a flare star and closer inspection using a faster chart speed was necessary to check this,' Hewish recalled.

October 1967

By October 1967 the last of the recorders and receivers had been installed and tests based on observations of the well-known source 3C273 had been completed. The IPS Array was now fully operational. Bell was now going out to the observatory daily to make fast recordings of the object at RA 1919, but to no avail. After the initial observations the source had vanished. For weeks on end Bell recorded nothing but receiver noise.[2] Hewish was anxious that the signal had gone, chiding Bell by saying it was a one-off event and that she had missed

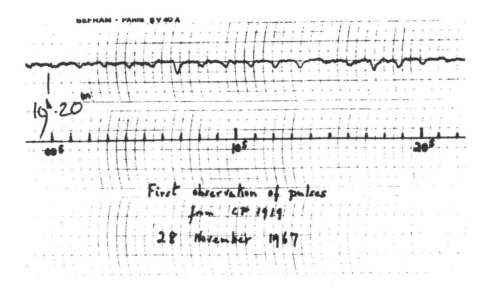

Figure 4b The discovery recording of CP1919. This fast recording revealed the telltale pulses that opened an entirely new field of astronomy.

it. Bell felt frustrated and so one day in late November 1967 she deliberately skipped her analysis to go to a lecture that sounded more interesting than her routine scrutiny of the recordings. Returning to the observatory the following day she scanned the recordings, and there was the scruff. Had Bell been at the observatory that night she would have seen it reappear. Encouraged that this was not a one-off event, Bell diligently continued her vigil and a few days later was rewarded with the return of the signal. This time she caught it on the fast recording. As the chart flowed under the pen it became apparent that the signals were more than a point source, they were a series of pulses. Taking the chart off the recorder, Bell spread the chart out on the floor and measured the spaces between the pen flicks with a ruler. The curious radio sources were flicking on and off with a period of 1.3 seconds.[3] It was November 27th 1967.

Bell contacted Hewish who was at the time teaching at an undergraduate laboratory in Cambridge. His initial reaction was that the source had to be artificial: there was no known natural phenomenon that could produce a recurring radio signal. Bell, on the other hand, didn't see why it couldn't be coming from a star. Nonetheless, Hewish was interested enough to go to the observatory at the time of the next transit. Despite the peek-a-boo game the source had played with Bell during those frustrating weeks in October, this time it performed on cue: right at the expected time, with Hewish and Bell looking on intently, the pulses reappeared.

Hewish checked the records, noting the time of day the source had appeared. The time it transited changed from night to night which showed it was keeping sidereal time, the time of the stars. The signal was coming from space, not from

Earth. They also ruled out alternative explanations for the signal. No, they weren't signals being reflected from the Moon, nor were they from an artificial satellite in a strange orbit. They even discounted the idea that they were the effects of a large, corrugated metal building to the south of the telescope! On closer examination the signal revealed that, whatever else, it was beyond the Solar System but within the Galaxy. Further, because the pulses were so evenly spaced the source could not be on a planet orbiting a star. If it was, the pulse period would change as the planet orbited the star, an effect known as Doppler shift. As the planet approaches us on one side of its orbit the pulses would be compressed and appear to have a shorter period; they would be more widely spaced as the planet receded on the other side. No such variation was seen, just a monotonous series of extremely well timed pulses. This was indeed a strange celestial phenomenon.

The next step was to verify once and for all that this new object was nothing to do with the IPS Array, and that meant observing it with an entirely different telescope with independent observers. Paul Scott and R. A. Collins were asked to do the verification using a different telescope. As the time of transit approached, Scott, Collins, Hewish and Bell crowded around the recorder. As they peered at the pens they saw ... nothing. The pens remained steady; no scruff appeared. Hewish, Bell and Scott wandered away in disappointment, but Scott continued watching the pens. Suddenly he cried out 'Here it is!' The others rushed back to the recorder and there was the signal. Due to a miscalculation of when the source was to transit the new site, the scientists had almost given up too soon. Here at last was the verification Hewish and Bell needed: the pulsating radio source was real. This was a radio source unlike anything previously observed. The questions now arose: who to tell, and how?

Meeting on How to Announce Discovery

You might think that making a new discovery like this is exciting and easy. Spot something new then rush to the newspapers. Science doesn't operate this way, however: it is a delicate balance of the urgency to get into the *scientific* press to gain priority and the need for verification. It was this last factor that was to stir the first controversy surrounding Hewish and Bell's discovery. In this case, they had discovered something new, but now they had to either explain it or show it was real. The difficulty with this new pulsing radio source was its contradictions. The rapid pulses meant the source had to be small. Stars are big, and while most vary periodically they do it over large periods of time, anywhere from a few days to months or years. At the extreme there is the constraint that nothing in the Universe travels faster than light. For example, even if the Sun were to suddenly stop shining for some unknown reason, it wouldn't wink out but rather fade gradually over a period of about a second. The reason is the diameter of the Sun is some 1.3 million kilometers and so the last rays of light from the farther side of the Sun would take about a second longer to reach us than the last rays from the nearer side. Because nothing can travel faster than light, the idea of something flickering on and off so rapidly implied it was very small. At the same time, the incredible accuracy of the period implied it was something large and regular. As Bell later put it, it was like a double decker bus turning around on a sixpence. It just didn't make sense. To make things worse, it occurred to the astronomers that such fast accurate pulses could also be artificially produced; the pulses even appeared at a frequency typical of signal generators. As we'll see shortly, artificial and celestial are a risky scientific combination. To avoid announcing a false discovery, the scientists had to be skeptical and they had to be thorough. That took a long time and hard work. But there was more to it than scientific diligence. To understand fully the reluctance of the Cambridge team to announce the discovery, we need to briefly go back a couple of decades.

Cosmological Conflict – Rivalry between Radiophysics and Cambridge

During the 1940s and 1950s, a rift had developed between the radio astronomy groups working at Cambridge and Sydney. The details of this unfortunate situation have been outlined elsewhere,[4] but the most important event that affected the Cambridge decision to withhold the pulsar discovery happened in 1957. Two years earlier Martin Ryle had presented, in the presence of Oxford University's Vice Chancellor, the cosmological implications of the Second Cambridge Catalog of Radio Sources of about 1,700 radio sources. At the time there was still controversy over the Steady State theory put forward by Fred Hoyle, Herman Bondi and Thomas Gold, and the Big Bang theory. The steady state theory proposes that the Universe has always been here and is never changing. In contrast the Big Bang theory says, as its name implies, that the Universe began at a finite time in the past in a single event and has been evolving ever since. Ryle showed that the 2C indicated a dramatic increase in the number of radio galaxies with distance, which contradicted the Steady State theory simply because greater distance meant looking further back into the past. If the Steady State theory were correct, there would be the same number of radio galaxies in the past (therefore seen at greater distances) as at the present (nearby) universe. Meanwhile in Australia the group led by Joseph Pawsey had also been counting radio galaxies. Based on a survey of 1,030 objects, the Sydney group found no significant difference between the number of galaxies observed and that predicted by the Steady State model. Despite the fact that the Steady State model was ultimately doomed, the important thing here is that the Australian results directly contradicted those presented by the Cambridge group. At stake was nothing less than the winning explanation for the origin of the Universe. In order to clear up the issue, Rudolph Minkowski from Mt. Wilson and Palomar Observatory traveled to Sydney and then Cambridge, spending weeks analysing both sets of data. He produced a long memorandum claiming the interpretation of the Cambridge data was indeed flawed, and that the Australians had got it right. In March 1957 the Australian group published a paper attacking the Cambridge survey. They pulled no punches, concluding with:

> The Cambridge survey is very seriously affected by instrumental effects which have a trivial influence on the Sydney results. We therefore conclude that discrepancies, in the main, reflect errors in the Cambridge catalogue, and accordingly deductions of cosmological interest derived from this analysis are without foundation.[5]

The Cambridge survey had lacked resolution, resulting in confusion between individual sources and weaker, multiple sources. The data was flawed and the Cambridge astronomers had tried to get too much information out of it. The subsequent 3C survey was much finer but still did not agree with the Australian results. There was still evidence of a significant excess of weak (more distant) sources which was contrary to Hoyle's theory. However, the damage to their reputation had been done. These remarkable scientists had made a mistake and paid the price. They weren't about to do it again. This time, they had to be absolutely sure these flickering, pulsing radio sources were real before they told the world.

Difficulty with Interpreting the Signals

A major difficulty with the pulsar discovery was its suspiciously artificial nature. All the facts pointed to a celestial origin, but celestial and artificial imply ... aliens. Decades later Carl Sagan would summarize this kind of scientific dilemma: 'Extraordinary claims require extraordinary evidence.' Shortly before Christmas of 1967 Bell walked in on a meeting between Hewish and Ryle on how to announce the discovery if Hewish's continuing Doppler shift measurements suggested a planetary origin. 'At that stage ETI (extraterrestrial intelligence) was still a serious possibility in my mind,' said Hewish. At the same time there was no proof that the signals were natural. Almost as a joke, Bell nicknamed the source LGM1 for 'Little Green Man One'. Supposing the signals were from an extraterrestrial intelligence, who were they to tell first? How do you make such an announcement? The discovery in fact made Bell very cross. 'Here was I trying to get a Ph.D. out of a new technique, and some silly lot of little green men had to choose my aerial and my frequency to communicate with us!' Undaunted, Bell returned to the laboratory after supper that evening to analyze more of the endless chart recordings. Then she found it: a patch of scruff at RA 11h 33m! She checked through the records and confirmed the signal. Although she had to leave the lab before it was locked up for the night, she checked that the source would be transiting the telescope later that night and determined to check out the second source.

When Bell arrived at the telescope in the early hours of the following morning it was very cold. Part of the system suffered a drastic loss of gain at low temperatures. 'But by flicking switches, swearing at it, breathing on it I got it to work properly for 5 minutes – the right 5 minutes on the right beam setting,' Bell recalls. 'This scruff too then showed itself to be a series of pulses, this time 1.2 seconds apart.' Here was a lifeline to Bell: it was extremely unlikely that two lots of aliens were transmitting to Earth at the same time and the same frequency but from different parts of the sky. Bell left the chart recordings showing the second source on Hewish's desk and left for Christmas in a much better mood.

Over Christmas, Hewish kept the survey going, replacing paper and ink and piling the paper chart recordings on Bell's desk for her to analyze. 'I continued the daily timing measurements for the vital Doppler information. I kept quiet until I had removed the planetary (ETI) possibility,' Hewish recalled. When Bell returned from the holiday she couldn't immediately find Hewish and so settled down to analyze more of the recordings. Now that she knew what she was looking for, it wasn't long before she discovered yet two more sources, one at RA 08h 34m and another about an hour away at RA 09h 50m. Within a few weeks, all three of the new sources had been confirmed. Bell waded through past records from the IPS Array, which by now amounted to several kilometers of paper chart. Although she found a few candidates there was nothing as positive as those first four sources.

The Announcement

In late 1967 rumours were circulating in the astronomical community of a spectacular discovery made by the Cambridge group. One member of the group working at Parkes, Max Komesaroff, would later recall that an 'informant' had told him the Cambridge group had been receiving 'signals from outer space'. Komesaroff was initially startled, but cooled as he realized that all of astronomy was receiving signals from outer space. In Komesaroff's words, when the news finally broke, 'it turned out that they had not in fact been receiving Morse-coded messages from alpha Centauri or from the Andromeda Nebula, but their story was only slightly less fantastic'.[6] But the Cambridge radio astronomers kept the news to themselves. Even at a regular Friday lunch organized to bring together scientists from a range of related disciplines at Cambridge, the radio astronomers went silent if the conversation edged towards what they were up to. Then in February 1968 it was announced that Antony Hewish would present a seminar on the discovery of a new type of pulsating radio source.

January – the Cambridge Seminar

On February 20th 1968, scientists began filing into the Maxwell Theatre in the Cavendish Laboratory. It seemed to Bell, sitting towards the rear, that every astronomer in Cambridge was there, including the controversial astronomer Fred Hoyle. Sitting expectantly on the wooden benches, they listened intently to

Hewish's announcement. Hewish revealed all ... well, almost all: his team had detected a rapidly pulsating radio source for which there was no immediate explanation. No one in the audience knew what the sources were and, despite the initial extraterrestrial intelligence speculation, it was now clear they had found a new astrophysical phenomenon. There was a feeling among those present that this was the dawn of something major in the world of astronomy. Hewish summarized the paper, mentioning that similar objects had been found elsewhere in the sky.

Discussion ensued over what these objects could be, with Hewish suggesting that they could be some kind of vibrating white dwarf. Fred Hoyle had other ideas. With his broad Yorkshire accent, he said he had not heard of these objects before, but that he thought rather than being white dwarfs, these new objects were supernova remnants. By this he didn't mean the expanding clouds of gas, but rather what was left behind: Baade and Zwicky's neutron star. Bell later recounted, 'Considering the hydrodynamics and neutrino opacity calculations he must have done in his head, that is a remarkable observation!'

In his autobiography *Astronomer by Chance*, Bernard Lovell recalled his reaction to hearing the news the day after the Cambridge seminar. Waiting for the start of a meeting of the Science Research Council in London, Lovell was joined by Hoyle who took the empty seat to his left. Hoyle had just returned from the US, and the hot topic that dominated astronomy at the time was quasars, especially using the large American telescopes to measure redshifts. Lovell asked Hoyle if there was any news. 'Not much about quasars,' he replied, 'but last night at a colloquium in Cambridge, Tony Hewish announced that he had discovered some radio sources which emitted in pulses with intervals of about a second.' Lovell was skeptical, and his immediate response was the signals must be interference. It was then that Hoyle told Lovell that the Cambridge team had been examining the sources for months, and that the evidence for an extraterrestrial origin of the pulses seemed convincing. Lovell returned to Jodrell the following day and 'was astonished to realize that, although our contacts with Cambridge were close, no one had any foreknowledge of this discovery [pulsating radio source]. Indeed it soon became clear that Hewish and his small group had for several months achieved a screen of security and secrecy about this work that was, in itself, remarkable in an age of instant communication between astronomers.'

A week later the paper announcing the discovery was published in *Nature*. Based on a mere three hours of observation, it ruled out the possibility of interference: these were a new astrophysical phenomenon. Still the Cambridge group kept the location of the sources to themselves for a further two months, a maneuver that annoyed and frustrated astronomers around the world eager to begin their own observations. It wasn't long before the location of the four sources was leaked, however. One of the astronomers working at Jodrell at the time was Francis Graham-Smith (who would in 1982 succeed Martin

Ryle as Astronomer Royal). 'When Lovell told me I telephoned Hewish, and pointed out that we could more easily follow up the four sources using the 250 ft telescope,' he recalled.'He said they had not yet completed the positional measurements of the three others, but within a day or so he let me have the positions and we were able to start observing.' The positions were also revealed to Allan Sandage at Palomar Observatory. Lovell leaked the positions to Hoyle, who passed the information on to Margaret Burbidge[7] then working at the University of California at San Diego. Soon observers and theorists the world over were studying the objects. The advantage that Jodrell Bank had over Cambridge was that whereas the Cambridge telescope could only monitor the pulsar for a few minutes each day as it passed through the telescope's beam, the giant telescope at Jodrell could follow the pulsar for as long as it was above the horizon. Within two weeks a second paper on pulsars was published in *Nature*.

Bell

Surprisingly, Bell ended her association with pulsars around this time, handing over the observations to the next generation of research students. She finished her chart recording analysis, measured the diameter of a number of radio sources, and wrote her doctoral thesis. The discovery of pulsars, which would bring fame and controversy, were included as an appendix. Although now recognized for her legitimate scientific contributions, including pulsars, the fame Bell initially received occupied the fine line between comedy and tragedy that only the popular press seem to be capable of. The *Nature* paper included a suggestion of a binary companion or planetary origin of the signals, which the media immediately took an interest in. Then they found out a woman was involved. A hundred years after Flemming, Maury, and Cannon had laboriously cataloged stellar spectra for their male master, newspaper journalists were asking this brilliant young scientist was she taller than or not quite as tall as Princess Margaret and how many boyfriends did she have? Bell recalls she was then photographed 'in several silly poses: standing on a bank, sitting on a bank, standing on a bank reading bogus records, sitting on a bank reading bogus records, running down a bank waving her arms in the air'.

Sadly, things haven't improved much. In 2004 Bell-Burnell pointed out that there is still a severe problem of a lack of female scientists.

Following the developments in pulsar research over the past 36 years has given me immense pleasure. More disappointing have been the developments in the recognition and advancement of women in astronomy. In December 2003, the International Astronomical Union (IAU) published an analysis of its membership. With only 10% of their membership female, the United Kingdom and the United States fall well below the international average. The only thing that has changed since a similar survey about 5 years ago is that the proportion in most countries has gone up a few percent. Admittedly, it tends to be the more senior astronomers who are IAU members, and there tend to be more women in the junior ranks, but at this rate it will take 50 years until 50% of senior astronomers are female. The pipeline is leaky: A higher proportion of females than males quit at each rung of the ladder.[8]

The media did make one positive contribution to the discovery, however. A science journalist from the *Daily Telegraph* asked the astronomers what they were going to call the new objects. While Hewish, Bell and the others had had more pressing issues to think about, the journalist suggested that, since they were pulsating radio stars, why not call them pulsars? The name stuck.

Nobel Prize Awarded for Discovery of Pulsars

But within the new science that Bell and Hewish had begun controversy and frustration was growing like a storm. It is important even in an account of the scientific developments of pulsars to acknowledge that the awarding of credit for scientific discoveries is sometimes not clearly defined, and the acknowledgment of the discoverer of pulsars is a fine example. The 1974 Nobel Prize for physics was awarded for 'pioneering research in radio astrophysics'. There were two recipients of the award that year: Martin Ryle 'for his observations and inventions, in particular of the aperture synthesis technique', and Antony Hewish 'for his decisive role in the discovery of pulsars'. There was no mention of Jocelyn Bell, and many have questioned this omission. At the time, however, one man was more vocal than all others: Fred Hoyle. A detailed account has been given of the episode in Simon Mitton's biography of Hoyle, but a brief sketch should give you an idea of what happened while conveying the concerns some had over the awarding of the Nobel Prize to Hewish.

Hoyle was in the United States delivering a series of lectures at various institutions. Hoyle was a very high profile scientist who was already deeply controversial within the scientific community and extremely popular with the public. It is no wonder that his comments, made to a newspaper journalist in Montreal, received a great deal of publicity. Following a press conference on quasars, he was asked what he thought about the awarding of the Nobel Prize to Hewish and Ryle. Hoyle responded with: 'Jocelyn Bell was the actual discoverer, not Hewish, who was her supervisor, so she should have been included.' This was bald criticism of the Nobel Foundation and tantamount to an accusation of academic dishonesty by Hewish. Hoyle went on to criticize the Cambridge group's initial reticence over the discovery. The media in typical style extrapolated Hoyle's comments, suggesting he had accused Hewish and his senior colleagues of stealing Bell's discovery. It wasn't long before Hoyle realized the seriousness of his remarks, and fretted to his friend Don Clayton from Rice University that, if Hewish decided to take legal action Hoyle could lose everything.

Naturally, Ryle and Hewish reacted immediately and sharply. Hewish couldn't see where Hoyle was coming from, and was angry and perplexed at such an accusation. It crossed Hewish's mind to take legal action, but in the end the issue was settled with a letter from Hoyle to the *The Times* clarifying his views, as well as a personal undertaking never to criticize the Cambridge group members again. In *The Times* letter he explained that he was impressed by Bell's participation in the discovery, that she had shown 'great persistence' in following up the seemingly spurious signal that turned out to be a pulsar.

> There has been a tendency to misunderstand the magnitude of Miss Bell's achievement, because it sounds so simple – just search and search through a great mass of records. The achievement came from a willingness to contemplate as a serious possibility a phenomenon that all past experience suggested was impossible. I have to go back in my mind to the discovery of radioactivity by Henri Becquerel for a comparable example of a scientific bolt from the blue.

Hoyle's opinion was that Bell was more than a mere worker, that she had shown genuine insight.

When I asked Hewish about the episode he pointed out that, particularly in the United States, there is a popular myth that Bell made the discovery single handedly while he hovered in the background. It's important to note that if it weren't for Hewish's long familiarity with radio telescopes, and his innovative design of an instrument that could detect rapid fluctuations in radio signals, pulsars would not have been discovered, at least not in 1967. Up to that time radio telescopes had been designed to smooth out radio signals to help deal with confusing and sporadic interference. It was the very nature of the IPS Array to detect rapid fluctuations in radio sources, and so in a way it was almost tailor-made to discover pulsars. In fact it would be months before any other pulsars were to be independently discovered by astronomers using a similar instrument in the Australian outback.

Hewish recalls that Bell was a 'diligent graduate student', but that she had no part in planning the sky survey of 'scintillating' quasars, or the design of the IPS Array. Her task was to keep the survey running and to analyze the recordings.

> Jocelyn did not have the experience to initiate further observations, such as accurate timing, distance estimates, or Doppler measurements to rule out the possibility of alien signals from an extra-solar planet! As University term had just ended I got on with this myself while Jocelyn re-examined past records for evidence of further possible pulsars (and came up with three more). I wrote the discovery paper having concluded that neutron stars were the most likely source of the pulses. So it was very much 'hands on' for me.

The Nobel committee are on record as having considered our respective contributions very carefully before making the award to me... I think a helpful analogy is the discovery of America, which is credited to Columbus rather than to the lookout who shouted from the masthead! In my Nobel lecture I gave full credit to Jocelyn for her contribution, and she was placed second author of our paper. Overall I think the Nobel award was appropriate.

What did Bell think of all this? Bell wrote in 1977:[9]

First, demarcation disputes between supervisor and student are always difficult, probably impossible to resolve. Secondly, it is the supervisor who has the final responsibility for the success or failure of the project. We hear of cases where a supervisor blames his student for a failure, but we know that it is largely the fault of the supervisor. It seems only fair to me that he should benefit from the successes, too. Thirdly, I believe it would demean Nobel Prizes if they were awarded to research students, except in very exceptional cases, and I do not believe this is one of them. Finally, I am not myself upset about it – after all, I am in good company, am I not!

Almost three decades later she wrote an important piece for the journal *Science*,[10] in which she commented:

Arguably, my student status and perhaps my gender were also my downfall with respect to the Nobel Prize, which was awarded to Professor Antony Hewish and Professor Martin Ryle. At that time, science was still perceived as being carried out by distinguished men leading teams of unrecognized minions who did their bidding and did not themselves contribute other than as instructed! Although I was not included, I celebrated that first award in 1974 of the Physics Prize for an astronomical discovery. Now I celebrate the fact that we have a better understanding of the teamwork necessary for scientific progress.

What is important is that a new astrophysical phenomenon was discovered that would open new windows onto the intricate clockwork mechanism of the Universe. The challenge to understand pulsars, however, had only just begun. Like their prediction, pulsars were about to attract the attention of some of the finest minds in science. As we will see, however, almost synonymous with brilliance is the word controversial. Pulsar astronomy was about to pick up speed.

References

1.   Right Ascension (RA) is the celestial equivalent of longitude. It is measured eastwards from a point on the celestial equator known as the First Point of

Aries, which is where the Sun crosses the celestial equator at the March equinox. RA is usually measured in hours, minutes, and seconds, with 24 hours being equivalent to a full circle.

2. It is now known that this was due to random interstellar scintillation.
3. For the perfectionist: 1.3372795 seconds. This more precise measurement was made later.
4. For example, see Robertson 1992, and Mitton 2005.
5. Quoted in Mitton, 2005.
6. Quoted in Robertson 1992, p. 305.
7. In 1972 Burdbidge became the first woman to become Director of the Royal Greenwich Observatory. However, she was denied the title of Astronomer Royal, which was given instead to Martin Ryle. There has never been a female Astronomer Royal.
8. *Science* 23 April 2004: Vol. 304. No. 5670, p. 489.
9. 'Little Green Men, White Dwarfs or Pulsars?' *Cosmic Search* Vol. 1, No. 1.
10. *Science* 23 April 2004: Vol. 304. No. 5670, p. 489.

5 'What makes pulsars tick?'

'New ideas in science are not always right just because they are new. Nor are the old ideas always wrong just because they are old. A critical attitude is clearly required of every scientist. But what is required is to be equally critical to the old ideas as to the new. Whenever the established ideas are accepted uncritically, but conflicting new evidence is brushed aside and not reported because it does not fit, then that particular science is in deep trouble – and it has happened quite often in the historical past. If we look over the history of science, there are very long periods when the uncritical acceptance of the established ideas was a real hindrance to the pursuit of the new. Our period is not going to be all that different in that respect, I regret to say.'

Thomas Gold
Journal of Science Exploration, Vol. 3, No. 2 1989

'I wish to point out that the oblique rotator model [of neutron stars] also results into an analogous emission of electromagnetic waves.'

Franco Pacini
Nature, Vol. 216, November 11, 1966

The late astronomer David Allen once described in his beautiful style the most remarkable feature of pulsars: how fast they spin. Imagine, he once said, that you were looking at a pulsar spinning on a pedestal. Paint a mark on the side of the pulsar near its equator. Now ask yourself: how fast would I have to drive in a car to keep up with that mark if the pulsar was spinning once a minute? Given a diameter of 20 km or so, you would in fact have to drive at a speed of around 3,700 kilometers an hour. However, pulsars do not spin once a minute, he went on. The first ones discovered spin in just a few seconds and it would not be long before even faster ones were found. It is no wonder that astronomers found such a concept difficult to accept, and for a long time the idea of pulsars being a spinning phenomenon was not taken seriously at all. No, there had to be another explanation.

Pulsar Conference: New York, 20–21 May 1968

And so it was that four months after the Cambridge announcement of the discovery of pulsars, their true nature remained an enigma. A conference was

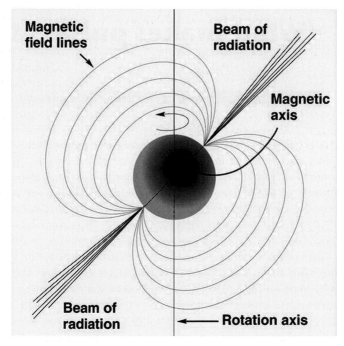

Figure 5 The rotating neutron star model that eventually explained the radio pulses. An off-set magnetic field produces radio beams, which are in turn swept through space by the rotation of the neutron star. When the beams sweep across Earth, it appears as a flash, or pulse, of radio waves. (Courtesy University of Arkansas at Little Rock).

organized to discuss the problem, and over two days during May 1968 invited astronomers from all over the world, including some of the original discovery group, met in New York to participate. The scientific organizer of the event was the Canadian-American astrophysicist, A. G. W. (Al) Cameron, a scientist who had spent years writing theoretical papers on the as-yet unconfirmed neutron star. Despite the grand aims of the conference and the calibre of the attendees, more than one astronomer who was around at the time described it as a 'fiasco'. Explanations centered on the concept of vibrating white dwarfs, an idea that was discredited before the year was out, some astronomers being led astray by false alarms of the detection of pulsars in visible light. But the most famous aspect of the conference was the rejection of the ultimately correct explanation of the nature of pulsars.

Vibrating Stars

Hewish had pointed out in the original discovery paper that 'The most significant feature to be accounted for is the extreme regularity of the pulses.' He went on to suggest that such regular pulsations are more likely due to some star-wide phenomenon rather than an isolated event in the atmosphere of the star. At the time of the conference there were in fact three widely considered explanations for the pulsating radio sources: oscillations, orbiting planets and rotation of the star itself. The most promising cause was oscillation, which is a regular and rapid vibrating back and forth. A common example of oscillation is a bell that has just been struck: the bell vibrates to produce sound. In a similar way, it was suggested that stars could oscillate releasing not waves of sound, but waves of electromagnetic energy that the astronomers were picking up as the radio pulsations. The rate at which a star could vibrate is based on the mass, gravity, radius and elasticity of the star. It was thought that ordinary stars couldn't oscillate like this, but there were two types of stars that might: white dwarfs and neutron stars. The previous year, Al Cameron had written a paper for *Nature* linking the then recently discovered celestial X-ray sources with Baade and Zwicky's still-hypothetical neutron stars and supernovae. The problem was that calculations had shown that neutron stars, if they even existed, would be oscillating up to 1,000 times faster, with periods of between 10 and 1 milliseconds, effectively ruling them out since nothing had been observed pulsing that fast. As far back as 1966 it had been shown that white dwarfs would naturally oscillate with periods of between 10 and 1 second. Despite the fact that the fastest pulsar known at the time of the conference had a period of a quarter of a second, considerably less than what the theoretical models allowed, white dwarfs were still within the realms of possibility and even regarded as the most likely candidate for pulsars.

Planets Orbiting Stars

There were alternatives to these explanations, however. One was that the radio pulses originated from a planet orbiting a star. Planets have shorter orbital periods the closer they are to their parent star. While it takes Earth 365 days to complete one orbit about the Sun, for example, the innermost planet of the Solar System, Mercury, can swing around the Sun in 88 days simply because it is closer, has less distance to travel, and has a greater orbital velocity. Then again, it needs this rapid orbit to resist falling into the gravitational well created by the Sun. In order to produce the rapid-fire radio pulses seen by astronomers, a planet would have to be orbiting a star with the same phenomenal periods: its 'year' would be measured in seconds! Such tremendous velocities implied very tiny orbits, so small that ordinary stars would quickly consume any planet in such close proximity. If a radio beam emitting planet were to follow such a tiny orbit around the Sun, for example, it would in fact be moving below the Sun's surface! Not likely. Even a white dwarf with the mass of a star compressed into Earth-like dimensions, 1,000 times smaller than the Sun, would be too big: any planet orbiting a white dwarf with a period of 1 second would have to graze the surface. It seemed more reasonable for a planet to be orbiting a much more compact neutron star, hundreds of times smaller than a white dwarf, but there were problems here as well. For one thing, the orbit would not be very stable. The energy of the system – planet and star – would steadily leak into space in the form of gravitational radiation, slowing the planet and causing it to spiral in to the neutron star within a few hours. Long before that happened, however, tidal forces would have ripped the planet to pieces: even a planet made of high tensile steel would need to be less than 20 meters across in order to withstand the titanic forces that close to a neutron star. Even assuming the planet could stay together and remain in orbit, what could possibly be generating the radio waves from the planet in the first place? No, there had to be another explanation.

Rotation

There was one idea that was not widely explored at the conference: rotation. Could the regular radio pulses observed from CP1919 and its kin be the result of something spinning? After all, rotation is one of the most regular and predictable phenomena in the Universe. From the stately rotation of the Sun and planets on their axes to the rotation of galaxies, spinning is a simple way to set up a repeating phenomenon. The question is what could be rotating so fast. The maximum speed with which a spherical object can rotate depends on mass and diameter. The smaller an object of given mass is, the higher the density, and the denser an object is, the faster it can spin without flinging itself apart. A good example to illustrate this is the planet Jupiter. This fantastic planet – 12 times the diameter of Earth and over 300 times its mass – spins on its axis in less than 10 hours. Because of its low density, however, it has an obviously flattened appearance like an under-inflated beach ball, a shape known as an ellipsoid, and if the planet were rotating even faster it would fly apart. If pulsars were in fact rotating on timescales measured in seconds, they would have to be made of sterner stuff than ordinary planets or stars in order to remain intact.

Once again, the most popular candidate was the small, massive, and very dense white dwarf. Understanding how a white dwarf could spin up to such speeds was not difficult either. White dwarfs are the collapsed cores of stars like the Sun. Now all stars rotate; in the case of the Sun it's about once a month. As we saw in Chapter 1, at the end of their lives the cores of stars begin to collapse, and as they do so their once stately stellar rotation is converted to a frantic spin. The classic comparison is with an ice skater initially spinning with arms extended, and then spinning faster and faster as she draws her arms closer to her body. Stars do the same thing at the end of their lives: as the core collapses to smaller diameters, the angular momentum is conserved and the core spins faster. White dwarfs, it was shown, can rotate with periods down to 1 second. At this speed, however, the white dwarf would turn into an ellipsoid, bulging at the equator and losing mass in a disk around its middle; any faster and it would fly apart. Even if it could remain intact at such speeds, a rotating white dwarf still did not explain the source of the radio pulses. If white dwarfs were not dense enough, there was only one other object that could do the trick: neutron stars.

Gold and Pacini's Rotating Neutron Stars

In Chapter 2 we saw how the discovery of the neutron led to the prediction of the existence of neutron stars. Neutron stars are quite massive – they can be anything from 0.2 to 2 solar masses – and yet they are tiny, perhaps 10 or 20 kilometers in diameter. Combined, these characteristics imply densities between 10^{14} and 10^{15} grams per cubic centimeter. In other words, a sugar cube of neutron star material would weigh in at 100,000,000 tonnes.

The rotating neutron star model was formally proposed independently by two astronomers, Thomas Gold and Franco Pacini. Pacini was the first to publish his ideas, which he wrote six months before pulsars were announced by Hewish and Bell.[1] He explained that neutron stars (remember, at this stage they were still hypothetical) would be the product of supernova explosions and would retain a great deal of energy in the form of heat, vibration, magnetism and rotation. The question Pacini posed was whether the release of this energy played any role in the excitation of supernova remnants like the Crab Nebula? (We will return to this extraordinary object later in the book.) Pacini dismissed heat and vibration as being significant sources of energy for supernova remnants, instead commenting that 'It seems more rewarding ... to look for some mechanisms by which the neutron star can release its magnetic or its rotational energy or both.'

It was assumed that neutron stars would be surrounded by intense magnetic fields. Ordinary stars have magnetic fields, too, but during the collapse of a massive star to form a neutron star the magnetic field would be compressed immensely. Now here is an important point: magnetic fields have a north and a south pole with the line joining them known as the magnetic axis. But the magnetic axes of planets and stars are typically offset from the rotation axis: as the star or planet rotates the magnetic axis appears to wobble in space. Pacini pointed out that the same would be true of the magnetic field surrounding a neutron star. Even if the alignment between the magnetic axis and the rotation axis was perfect at the beginning of the collapse, supernovae are not exactly precision events, and it is unlikely that the two axes would remain aligned during the star's collapse. Therefore, as the neutron star rotates at its fantastic speeds, the magnetic axis gyrates at the same frantic pace. Fred Hoyle, Jayant Narliker and John Wheeler had already suggested that radio waves would be emitted due to the vibration of a magnetic neutron star, but Pacini pointed out the vibrations would quickly subside, casting doubt on this explanation. As an alternative, Pacini suggested that the 'oblique rotator model' might also emit electromagnetic radiation. Further, the radio waves would be emitted with the same frequency as the rotation of the neutron star. Unfortunately, Pacini's paper seems to have gone unnoticed by most observers, and even Thomas Gold was unaware of Pacini's ideas. Ironically, the two astronomers were working in almost adjacent offices at Cornell University when Gold wrote his paper linking pulsars with neutron stars, but for whatever reason they did not cross theoretical paths. At least, not then.

Gold's Explanation of Pulsars

At the time of the pulsar discovery Gold was Director of the largest radio telescope in the world, the giant 300 meter radio telescope at Arecibo, on the island of Puerto Rico in the Caribbean Sea. Gold was a controversial figure to say the least. From the very start he had a knack of offering theories in branches of science about as divergent as you can get. Just a few of his ideas will serve to illustrate just how diverse was his thinking. As a graduate student in 1946, he spotted an error in an accepted theory of hearing, and offered an alternative theory that the inner ear acts as a kind of amplifier. His theory was confirmed three decades later when physiologists discovered the tiny hairs in the inner ear that do just that. In 1948, he leapt from physiology to cosmology and teamed up with Fred Hoyle and Herman Bondi to promote the (now discredited) Steady State theory of the origin of the Universe. He even predicted that the Moon would be covered by a thin layer of dust, something confirmed by Neil Armstrong's 'one small step'. By the end of his life he was theorizing on where all the Earth's oil reserves come from. Rather than being the result of accumulated deceased organisms, Gold proposed that oil was produced by primordial hydrocarbons dating back to the origin of the planet itself. This same extraordinary scientist came up with the rotating neutron star explanation of pulsars.

Gold submitted a paper describing his theory to the New York conference, but it was rejected. Al Cameron was no stranger to neutron stars, having written numerous papers about their properties for years despite neutron stars remaining hypothetical entities. In fact, he had suggested a link between neutron stars and X-ray sources three years earlier. Yet after he had read Gold's paper linking pulsars with neutron stars he refused to give Gold even five minutes of time at the conference. Gold's suggestion was, Cameron said, 'so outlandish that if this were admitted there would be no end to the number of suggestions that would equally have to be allowed'. Fortunately, the Editor of *Nature* at the time, John Maddox, had a knack for seeing the scientific value of new ideas and accepted Gold's paper on the subject. It took less than a week for the paper to appear in print following its arrival at the *Nature* offices, possibly a record for a scientific publication! One of the most important parts of the paper is the last paragraph before the credits. Here Gold predicts two things. First, pulsars should be very gradually slowing down. Second, if they were slowing down, then, by rights, there should be younger, faster pulsars waiting to be discovered. Gold didn't have to wait long for both of these predictions to be spectacularly confirmed.[2]

Vaughan, Large and the Vela Pulsar

A newly completed radio telescope in the Australian outback was about to change the game forever. In 1968 Alan Vaughan, now senior lecturer at Macquarie University in Sydney, had begun his PhD under the supervision of

Michael Large the same week the paper announcing the discovery of pulsars was published. One of the other faculty members, Tony Turtle, asked Vaughan if he wanted to join him at the Molonglo Radio Observatory near Canberra searching for the four recently announced pulsating radio sources. The primary instrument at Molonglo was the Mills Cross, a pair of mile-long, narrow radio antennas forming a cross that produced a single beam. It was designed by Bernard Mills from Sydney University who had earlier constructed a similar instrument near Sydney. The Mills Cross was opened in November 1965 with the goal of providing a detailed survey of radio sources. Naturally, Vaughan leapt at the opportunity and together with Turtle used the Mills Cross to observe the four pulsars. They then began to wonder how they could use the telescope to discover more.

The Mills Cross telescope had a wide field of view and so could scan large areas of sky and find pulsars whose positions were not well known. Initially Vaughan and Turtle 'split' the east-west arm into an 'east arm' and a 'west arm' by tilting them slightly differently so they were separated by one or two degrees declination to increase the amount of sky covered using the transit method. This way, as Earth rotated, they were scanning large areas of sky in declination. In the meantime, Large and Vaughan discussed a project that would improve the sensitivity of the Mills Cross. Up to that point, the east-west arm of the Mills Cross produced a single beam. The device Large and Vaughan had in mind would introduce a phase delay in two directions across the east-west arm producing two extra beams, which were independent of the central beam: one would be a bit ahead of it in time and one was a bit behind. When the signals were combined with the signal from the north-south arm the result was three independent observations of the source as it transited, which significantly improved the sensitivity of the telescope. It was Vaughan's job to build a device to do this.

Vaughan and Turtle started going to the telescope every few weeks, using the telescope to search for pulsars for a week or so each time. It wasn't long before they discovered the first two southern pulsars, one of which had a ~0.25 second period, telling them they had a fairly good sensitivity to pulsar periods. Turtle moved overseas shortly after these discoveries, leaving Vaughan and Large to continue the search. It was then they realized that having three sequential beams in the east-west arm – initially devised to improve the sensitivity of the telescope for all radio sources – was actually a good way to search specifically for pulsars. A major and increasing problem for all telescopes is interference, and radio telescopes are no exception. The trick for Vaughan and his colleagues was being able to distinguish between the relatively weak pulsar signals and background radio interference. Such interference can come from sources as wide ranging as distant thunderstorms to faulty ignition systems in passing cars. The key difference is that interference comes from Earth, whereas pulsars are decidedly a celestial phenomenon. At that stage, the east-west arm was recording simultaneous transits of each source and so didn't discriminate against interference. But by having an early and late beam the astronomers could distinguish between a celestial point source of radio waves and background interference: while interference was recorded simultaneously on both beams, a pulsar would transit first through the early beam and then through the late beam. The signal from each beam was recorded on chart paper by two different pens, so if a regular train of pulses was traced out by one pen (recording a pulsar's transit through one beam) and then later by the other, the astronomers knew they had a pulsar. This became the secret of the first detailed survey of the southern sky using the Molonglo telescope.

Vela Pulsar Discovery

It was during this survey that Large and Vaughan made their major discovery. One weekend, Vaughan was off at a church camp, with Large in charge of observations. Coincidentally, Large had just begun an observing run one morning and so was watching the chart recorder when suddenly it went crazy. First one pen began scribbling madly on the paper, and then a short time later so did the other. Clearly, whatever this was, it was celestial. Large knew something dramatic had just happened and so immediately contacted Vaughan, who got there as quickly as possible. The pair knew they had found a pulsar, and also that it was fast … too fast in fact for their chart recorder. Large also contacted Bernie Mills, who checked the position of the pulsar and found something even more interesting: it was close to the middle of the Vela supernova remnant, itself the result of a star that exploded some 10,000 years ago. It was at this stage they first realized that what they had found was more significant than yet another pulsar. As the astronomers later pointed out in their paper in *Nature*, this could be a coincidence. However, all the facts – location and distance – placed it close to the supernova remnant. It was not only possible but likely that the pulsar was the remains of the supernova just as had been predicted by Baade and Zwicky nearly four decades earlier.

For the next observation, they ran the chart recorder as fast as they could, but still the pulses came so fast the pen simply blurred them into a single mass as the pulsar passed through the telescope's beam. In order to distinguish individual pulses and determine the pulsar period, they had to come up with a new strategy that involved some low but very effective technology. They disengaged the mechanism of the chart recorder so that the paper could be pulled freely under the pens. Disconnecting one of the pens from the telescope, they attached the pen to the mains, which operate at 50 Hz. Now, instead of flicking in time with the pulsar, it flicked 50 times a second across the paper providing an accurate time source. The other pen remained connected to the telescope. Just before the time of transit, Large or Vaughan would take hold of the free end of the chart recording. As soon as the pulsar passed into the beam of the telescope, whoever had the chart would run down the corridor dragging the paper behind them. As one pen recorded the 50 Hz from the mains as an accurate time source, the other pen recorded the pulsar signal alongside it, allowing them to determine the frequency of the pulsar. Using this technique, they found the Vela pulsar was flashing on and off 11 times a second, a new pulsar record. (The astronomers eventually obtained a fast chart recorder and obtained more accurate results.)

The Vela pulsar was the fastest so far discovered, but it wasn't long before astronomers found an even faster one. A month later the announcement was made that astronomers using the 300 meter Arecibo telescope in Puerto Rico had discovered an even faster pulsar in the vicinity of the Crab Nebula, another supernova remnant. We will be returning to look more closely at the major role the Crab has played in the story of pulsars in the next two chapters. For now, we will look at the evolution of the modern understanding of pulsars that was confirmed with the discoveries surrounding the Vela pulsar.

Parkes and the Vela Pulsar

In late 1968, the 64 meter Parkes radio telescope in western New South Wales centered on Large and Vaughan's position of the Vela pulsar to take a closer look. Parkes had played an important role in quasar research (see Chapter 3), and in the years to follow was destined to become the source of more pulsar discoveries than any other observatory in the world. Parkes astronomers had contributed to pulsar astronomy almost from the word go. As recounted in Haynes, Haynes, Malin and McGee's comprehensive history of Australian astronomy 'Explorers of the Southern Sky', the Cambridge announcement of the pulsar discovery caused a great deal of excitement in the Radiophysics Laboratory which operated the Parkes telescope. Two weeks later a group of astronomers was sitting in the Parkes control room discussing how best to conduct a search for more pulsars. Meanwhile, the telescope driver searched for CP 1919. One of those present was Brian Robinson who recalled he 'heard some unusual movement of the pen chart recorder'. The pulses were at first not resolved, so Robinson increased the speed of the recorder and the well-known pulse pattern emerged. Before its replacement with plastic notes, the original paper Australian $50 note preserved this recording of CP 1919, along with an image of the Parkes telescope. In March 1968 Venkataraman ('Rad') Radhakrishnan, Max Komesaroff and David Cooke used the Parkes telescope to observe CP1919. For four days, they were able to monitor the pulsar for as long as it was above the horizon. While the Cambridge group was limited to observing CP 1919 for a few minutes each day as the pulsar passed overhead, the fully steerable Parkes telescope was able to track the pulsar as it moved across the sky hence recording the pulses continuously. They found that the period of the pulsar was a little different from that reported by Hewish. Apparently, the British team had miscounted the number of pulses by 1, resulting in a tiny period error of 0.000 02 seconds. The Australians quite properly reported the Cambridge team's error. The precise reaction of the British is not clear, but one imagines that the Cambridge group was not pleased.

Flips and Slow Downs

When Parkes was used in order to study the Vela pulsar, an important discovery was made that would help settle the issue of the nature of pulsars: the radio waves emitted by the pulsar are polarized. As a wave of light is traveling through space, it vibrates perpendicular to the direction of travel. In non-polarized light, such as the light from the Sun, the light is vibrating in all directions. When light is polarized, however, it vibrates in one direction only. This can happen when light reflects off non-metallic surfaces such as glass or water, and the result is an intense glare seen in, for example, the reflection of sunlight from car windows. The radio beams from pulsars, it turns out, are also polarized, and in December that year, Radhakrishnan discovered that the polarization angle of the radio beam from the Vela pulsar flipped over during each pulse.

Radhakrishnan pointed out this would make sense if the beams were coming from a magnetic pole that was offset from the rotation pole of the neutron star, just as Gold and Pacini had predicted. He went further, however, by showing that a rotating neutron star accounted not only for the observed periods: the radio source could also be explained by an immensely collapsed and concentrated magnetic field. When a neutron star forms, not only is the stuff of which it is made compressed, the magnetic field it possessed as a normal star collapses with it and is intensified enormously so that it becomes a million million times stronger than the Earth's magnetic field. Electrically charged particles near the pulsar would travel along the magnetic field lines and be accelerated enormously all the way to relativistic speeds (i.e., close to the speed of light) as they follow the curve of the lines. These particles would in turn produce a beam of radio waves radiating along the magnetic poles of the neutron star. As the pulsar rotates, the beam from the magnetic poles would sweep through space like the beam from a lighthouse. If the beam swept across Earth and its many radio telescopes, what would be seen is a flash of radio waves with each rotation of the neutron star. The theory became known as the 'polar cap model' and was followed up by more theoretical work by Max Komesaroff. Although the actual way the radio beams are produced is still not fully understood, the idea of emissions from the magnetic poles of pulsars is still the accepted model.

Slow Down

One of the predictions made by Gold and Pacini is that pulsars slow down as they get older: the energy needed to produce the powerful radio beams came from the loss of rotational energy. This slow down was confirmed by Radhakrishnan and Dick Manchester, then a junior scientist at Parkes, using the Parkes telescope. Comparing observations of the Vela pulsar's period in February 1969 with that measured the previous December, they discovered that the period had decreased by 10 nanoseconds per day. Two weeks later Radhakrishnan and Manchester were again using the Parkes telescope to observe the Vela pulsar. By this time, they thought they knew the pulsar reasonably well and set up the equipment at the 'known' pulse frequency. To their surprise, 'the pulse began marching backward across the screen'.[3] The pair spent some time checking the equipment to see what they had done wrong, and eventually Radhakrishnan went to bed leaving the younger Manchester to sort out the problem. After the Vela pulsar had set, Manchester observed a number of other pulsars, which were all behaving normally. They convinced themselves that in fact there was nothing wrong with the equipment, but rather that the Vela pulsar had sped up by one part in a million. The pulsar 'glitch' was independently observed by P E Reichley and George Downs using the 70 meter dish of the Jet Propulsion Laboratory's Goldstone Tracking Station. Papers announcing the discovery were published simultaneously in the same issue of *Nature*. Exactly what causes these glitches remains a mystery, but may be linked to changes in the superfluid interior of the neutron star. Herein lies the importance of the discovery: it may be that studying glitches is one of the few ways scientists can study the behavior of ultra-dense matter.

Where To From Here?

By the early 1970s, astronomers were scanning the skies in search of more pulsars, a search that revealed a zoo of bizarre objects and phenomena, each as unexpected as the original discovery. But before these surveys began, there was one more pulsar to be discovered in a non-systematic manner, and this was to play a leading role in pulsar astronomy in the years to come. It all began with the death of a massive star in the year 1054.

NAMING PULSARS

Vaughan and Tony Turtle chose the now universally accepted 'PSR' as an abbreviation for pulsar. Until then pulsar names had been a combination of the abbreviations of the observatory that discovered them and the RA of the pulsar. Hence, the original pulsar was called CP 1919, the CP standing for 'Cambridge Pulsar'. 'We decided we'd apply PSR to pulsars whose positions were accurately known, whereas inaccurately known pulsars were called MP for Molonglo Pulsar,' Vaughan told me. They decided that once a pulsar's position was accurately known it didn't belong to them anymore – anyone could find it – and so it became PSR, short for pulsar. 'We tried PRS for 'pulsing radio source' and things like that, but PSR had a nice ring to it, and it's been used ever since.'

References

1.   The paper was published in the November 11th 1967 issue of *Nature* , but written about six months before the announcement by Hewish *et al* in February 1968.
2.   It should be mentioned that apparently Gold's predictions were not unique. One astronomer I contacted said many were amazed when Gold received credit for the rotating neutron star model, since it 'was a common piece of coffee time speculation'.
3.   Manchester, D. *'Pulsars at Parkes – Past and Present.'*

6 'The Crab'

The discovery of the Crab and Vela pulsars put astronomers on the right track in deciding what pulsars were. The Crab pulsar, however, was to play an even more important role in pulsar astronomy. Aside from the intrinsic scientific interest of this object, it lies at the center of two stories of scientific priority. Priority – who discovered something first – is a big issue in science. Major discoveries always rely on technology (and hence technologists) and generally follow on from the work of others. Yet the fact remains that the scientist who gets their work into print earliest is given the credit for the discovery. An important and recurring factor in all this is not only making the right observations at the right time, but recognizing what you've seen. There are many examples of this in the story of pulsars, and perhaps none is more dramatic than the discovery of pulsars themselves. While Bell and Hewish have properly gone down in history as the discoverers of pulsars, pulsars were first detected, but not recognized, much earlier. Using an X-ray telescope, American astronomers had detected the pulsating signal from the Crab Nebula months prior to Hewish and Bell's discovery; the trouble was, no one expected to find pulses in the observations and so, at least at first, no one thought to look. We'll return to that story later, but first it is worthwhile to take a closer look at the discovery of the Crab pulsar, a discovery that unfolded in two episodes.

Clues to the Crab Pulsar

The Crab Nebula was known to be an unusual object. Lying at the site of a supernova observed nine centuries earlier, it was still shining despite the fact it should have faded long ago. Something was powering this object, and it was likely to be an unusual star lying in the center of the nebula. This star, the south western star of a pair in the middle of the nebula, was an intense, blue colored star with a spectrum that showed none of the usual spectral lines associated with normal stars. The nebula itself was a strong radio and X-ray source, further signs of the fact that such a prodigious energy output should have dimmed the nebula. In 1964 Tony Hewish and Samuel Okoye had discovered a compact radio source near the center of, and yet separate from, the radio emissions of the nebula itself. Further, they showed that the radio source was scintillating, indicating that whatever was emitting the radio waves was small. It is against this backdrop that the search for a pulsar in the Crab Nebula was played out.

Discovery of the Crab – Dave Staelin and Ted Reifenstein

At the time of the announcement of pulsars, Dave Staelin was a new faculty member of the Massachusetts Institute of Technology. Having been trained at the same institution, his department head told him it was imperative he took a leave of absence and spend time somewhere else. To Staelin, the National Radio Astronomy Observatory's Greenbank Telescope in West Virginia seemed like the perfect place. But what to do? Staelin wanted to combine his interests in signal processing and radio astronomy, and the newly discovered pulsars seemed the logical choice. By the time he arrived at Greenbank in June 1968, Staelin had already developed a concept of a system that could simultaneously detect not only the periodicity characteristics of pulsars, but also a characteristic called dispersion. Radio waves – simply another part of the electromagnetic spectrum – can be dispersed just like light. As radio waves pass through interstellar space they encounter free electrons that delay waves with longer wavelengths more than those with shorter wavelengths. Now the thing about pulsars is that they don't transmit their pulses all in one wavelength, but rather over a range of wavelengths. So as a single pulse passes through space, the longer wavelengths will be delayed more than the shorter ones by the time they reach Earth. By looking at how much of a difference there is between the arrival times of a single pulse at two different wavelengths, astronomers can tell how much the pulse has been dispersed. This is a way of determining how much space the pulse has traveled through, and hence how far away it is.[1]

Within months Staelin and Edward (Ted) Reifenstein, another recent arrival at the National Radio Astronomy Observatory (NRAO), and engineers Bill Brundage and K. A. Braly had developed the spectrometer and software needed to make sense of the signals. The equipment was installed at the 91 meter Greenbank radio telescope. This telescope has since been replaced by a newer instrument following the original's demise in 1988.[2] Just like Hewish's IPS Array, the original Greenbank telescope couldn't point just anywhere in the sky: although an independently mounted dish it was only able to nod up and down along the north south line and used the Earth's rotation to scan in Right Ascension. By shifting the elevation of the telescope a little once every 24 hours, the astronomers were able to scan the entire sky.

Searching

Staelin and Reifenstein were interested in searching the plane of the Galaxy for pulsars for the simple reason that, if pulsars were the result of supernova explosions, this would be the most likely place to find them. Supernovae are the result of massive stars reaching the end of their lives. Since massive stars don't live more than a few tens of millions of years, it was reasoned that they wouldn't have had time to drift far from the plane of the Galaxy where the vast majority of star formation occurs. Although the Crab Nebula fell just outside their search area, Staelin and Reifenstein asked the telescope time allocation committee for permission to search the area of the Crab as well. Their request was honored.

The astronomers had use of the telescope for half a day each day for a couple of weeks and took turns at the telescope along with an operator. The observations were recorded on magnetic tape and returned to Charlottesville, Virginia, where they were processed at the central National Radio Astronomy Observatory's computer laboratory. The astronomers ran the IBM 360 computer at nights: there was no budget for computer operators, and the daytime runs were allocated mostly to others. The computer program was divided into two main parts, a preliminary filter matched for each combination of pulsar period and dispersion of interest, and a second pass that searched more intensely the most promising candidates. The candidates were scored by both period and dispersion.

Irregular Spikes

In order to study the dispersion of the suspect objects, their signals were recorded at 50 different wavelengths, or channels. Staelin printed out the data for these candidate pulsars as a threshold plot over the 50 frequency channels they had observed, indicating the signal strength in each of 50 columns with an asterix or other character. 'That night I took home many inches of printout and was startled to discover very strong dispersed pulses crossing the page – all 50 channels – at random intervals of a few seconds or more,' recalls Staelin. The random nature of the pulses didn't rule out a pulsar. You may remember that when Bell detected the first pulsar, it was only possible because Hewish had designed the IPS Array to respond to rapid fluctuations in the radio signals. If the detectors sample a signal less frequently than the actual pulse rate, then the source appears steady. Staelin and Reifenstein's equipment sampled the signal every 60 milliseconds, a far shorter time than the fastest pulsar so far known, and still the signal appeared only as a random spike. In mid-October, a series of strong, sharp radio signals were detected, but the signals occurred at random intervals unlike the clockwork nature of the pulsars then known. Was this a celestial signal or noise from somewhere in the system? Although the signal didn't have the regular ticks seen in other pulsars, it did peak at discrete times almost as if the telescope were picking up random giant pulses from a steady and amazingly fast sequence of pulses. This had to be a pulsar.

When Staelin and Reifenstein looked closer at the data it became clear that there were in fact two sources – NP 0532 and NP 0527 – very close together in the sky. Because the beam from the telescope was wide, the two signals had overlapped slightly; what distinguished them were their dispersions, which were different, and told the astronomers they had found not one but two pulsars. One of the pulsars, NP 0532, stood out: its position coincided with the Crab Nebula supernova remnant. If this was a pulsar, it would be further vindication of Baade and Zwicky's prediction of the neutron star-supernova association: the possibility of the pulsar-supernova remnant being a mere chance alignment

Figure 6a Crab Nebula. STScI-2005-37. (Courtesy NASA, ESA, Jeff Hester and A. Loll (Arizona State University).)

was vanishingly small, and so this pulsar suspect would further link pulsars to neutron stars. Further, it would be confirmation of Gold's prediction of a pulsar in the Crab Nebula.

Follow-up observations pinned down the location,[3] but the sampling time was too long to determine its period. The best they could do was put an upper limit on the period of 0.25 seconds, its shortest observed inter-pulse separation.[4] Staelin and Reifenstein briefly considered keeping their discovery secret until they could learn more, but quickly decided to publish the dual discovery without knowing the periods exactly: they reasoned that sharp, highly dispersed pulses could only come from a pulsar, regardless of period. Nonetheless, random pulses do not a pulsar make, and the publication of the discovery of even random pulses from the Crab Nebula – a supernova remnant – spurred astronomers into action. In particular, a team of astronomers not only confirmed that there was indeed a pulsar in the heart of the Crab Nebula, but determined its astonishing period and pinpointed its location. They did this using the largest telescope in the world, the Arecibo radio telescope.

Arecibo and the Crab Pulsar

The Arecibo radio telescope is a stupendous instrument. Operated by the National Astronomy and Ionospheric Center of Cornell University, it was first brought into operation in 1963. At 305 meters from side to side, it is the largest single radio dish on the planet. It is impossible to mount such a dish in the conventional, fully steerable manner of other telescopes. Instead, the wire mesh dish[5] was built into a natural depression in the hills of Puerto Rico near the town of Arecibo. It stares perpetually toward the zenith and uses the Earth's rotation to allow it to scan the sky. Suspended 130 meters above the dish is an enormous receiver complex, as long as a football field, and capable of pointing at different parts of the dish, allowing the telescope to 'point' at different places on the sky within 20 degrees of the zenith. Although this limits its range somewhat, one of the objects that crossed its gaze was the Crab Nebula.

In April 1968 Richard Lovelace was a graduate student at Cornell University under the supervision of Professor Edwin Salpeter. Salpeter was very interested in pulsars and so sent Lovelace to Arecibo to do research into pulsars. There he was joined by other students, including David Richards, John Comella, Hal Craft and John Sutton. The young astronomers weren't short of support from management: the Director of the observatory at the time just happened to be Tommy Gold and although it can't be said he initiated the Crab pulsar work he was very supportive. The on-site Director of the telescope itself was Frank Drake, a scientist who had pioneered what would later become known as the Search for Extraterrestrial Intelligence (SETI) in the early 1960s.[6] He also actively pushed the search for new pulsars. Although there were telescope operators on hand at Arecibo, in the late 1960s, individuals, including graduate students, were allowed to run everything, including the telescope steering, the radio receivers and the computer. There was a continual problem of breakdowns of different things: even though the receivers and computers were in an air-conditioned building the tropical air found its way in. Also, the Puerto Rican electrical power was terrible with the frequencies wandering all over from 55 to 65 Hz, and sometimes one of the phases of the three phase power would drop out.

1968 marked the time when analysis of radio telescope data went from analysing paper chart recordings to using computers. As we saw in the last chapter, Vaughan and Large used a somewhat unusual technique to refine their observations of the Vela pulsar, but this would be one of the last times a major pulsar discovery would be made without computers. It was Lovelace's job at

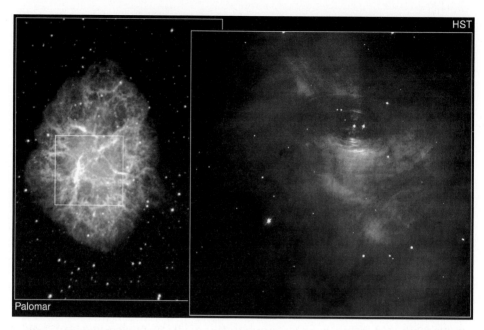

Figure 6b Crab Nebula in close up. This extraordinary image reveals the pulsar (left of the two stars) at the heart of the Crab Nebula. STScI-1996-22 (Courtesy NASA, ESA, and Jeff Hester and Paul Scowen (Arizona State University).)

Arecibo to develop computer codes for searching for pulsars using the observatory's CDC 3200 computer, and then to analyze the data collected by the telescope for pulsar signals. In October he began working on a new computer code specifically designed to locate weak periodic signals from radio 'noise'. 'What was notable about our work at that time was we had the expectation of rather high frequencies (50 Hz) corresponding to rapidly rotating pulsars,' Lovelace recalls. 'But at that time many scientists thought that the objects were white dwarfs which could give only low frequencies (about a hertz or so).' On one of Tommy Gold's visits to the Observatory in October, Lovelace recalls that Tommy strongly encouraged him to look for high frequencies expected for rotating neutron stars. Lovelace's new program was called 'Gallop' because it was so fast at doing Fourier transforms of long (about 16,000) data samples. Lovelace had access to the computers between midnight and 8am during the week and all day on weekends. On the night of November 9/10 1968, he discovered the first four digits of the Crab pulsar's period, 33.09 milliseconds. Soon after this, the period was greatly refined by David Richards who also found that the period was gradually increasing. The position of the Crab pulsar was initially narrowed down by allowing the pulsar to drift across the telescope's field of view and observing its strength at 195 MHz with the Gallop code. About a week later observations at 430 MHz were able to pin down the location of the pulsar to within 5 arc minutes of the center of the Crab Nebula, the accuracy of this observation being so important to the subsequent optical discovery discussed in the next chapter.

Giant Pulses

The importance of the giant pulses should not be understated in the discovery not only of the Crab pulsar, but of the realization of the existence of pulsars with short periods. Dave Staelin told me that others had also examined the Crab seeking periodic pulsar emission, but most instruments did not sample fast enough or make use of computers to detect short periods. 'We were successful because of the 'giant pulse' phenomenon, wherein the Crab occasionally emits pulses of enormous strength separated by much weaker periods of emission lasting seconds or more, but because of their rare character and non-periodicity these giant pulses are difficult to identify manually or any other way. Repetitive dispersion was their revealing signature in our case.' It is also important to highlight that were it not for the fact that the Crab pulsar emits such giant pulses, pulsars of such astonishingly short periods might not have been discovered by radio astronomers at all. None of the general radio surveys carried out over the next 16 years would have detected the Crab or any other pulsar with such a short period. The chances are astronomers working at other wavelengths, particularly optical and X-ray wavelengths, were recording observations with sufficient time resolution to successfully discover the Crab. As we'll see in the next chapter, the Crab pulsar was the first to be detected in visible light, itself a story riddled with the controversy of priority. But before we leave the Crab pulsar, let's see just how the announcement of the first radio pulsars led to the realization that pulsars had already been detected but unrecognized by the time they were discovered.

They Were in the Data All the Time

A month before Jocelyn Bell began her historical observations that led to the discovery of pulsars, a young American graduate student, Gerry Fishman, gazed into the clear predawn sky above Palestine,[7] Texas. Above him rose a 130 meter diameter balloon carrying a unique astronomical instrument. Unlike radio or optical telescopes, this one was capable of observing the Universe in γ-rays (gamma rays) and X-rays,[8] two forms of high energy radiation that represent, not surprisingly, some of the most powerful phenomena in the Universe. The problem is that Earth's atmosphere filters out γ-rays and X-rays. Good for us, but lousy if you want to observe the high energy universe. To do that, you need to get above the atmosphere. Hanging beneath a balloon seems an unlikely place for a telescope, but short of using space-borne instruments this was the best option.

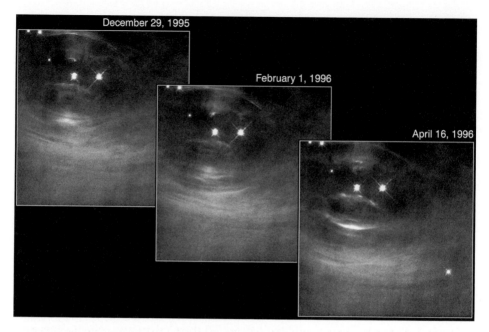

December 29, 1995

February 1, 1996

April 16, 1996

Figure 6c Changes in the structure in the heart of the Crab Nebula are visible in this sequence of images spanning three and a half months. (Courtesy NASA, ESA, and Jeff Hester and Paul Scowen.)

Because the telescope had to be in its lofty position to observe the Crab later that day, the balloon had to be launched before dawn, and Fishman and his colleagues had been up most of the night preparing the instrument. 'The tremendous size of the balloon, over 50 stories high, along with the jet-engine like noise during the helium filling process, and the large launch pad to accompany the special large launch vehicle which was able to hold the gondola, maneuver and be able to accommodate last minute changes in wind direction, all combined to give a surreal appearance to the scene,' Fishman recalls. 'Local residents always showed up for a balloon launch, along with some members of the press. For our launch, friends and colleagues from Houston made the four-hour drive to Palestine.' As Fishman watched, the balloon climbed to almost 40 kilometers above the surface of the Earth, 99.5% of the atmosphere was beneath it. Guided by remote control, the telescope swung around and pointed at its target: the Crab Nebula.

Gerry Fishman

Fishman had entered graduate school in the new Department of Space Science at Rice University in Houston, Texas in 1965, and began to explore which research specialties existed in Space Science. One of these was gamma-ray astronomy. He went to the library to try to discover more about the field, but hardly anything had been written about it. Sensing that this would be the chance to get in on the ground floor of a new and potentially exciting field, Fishman joined the gamma-ray astronomy group, headed by Robert C. Haymes, that summer, before classes began.

The Rice University group was one of only a handful of such groups in the world performing balloon-borne observations of high-energy objects. Two other groups performing similar research were from MIT headed by George Clark, and from the University of California, San Diego headed by Laurence Peterson. These three groups were in a highly competitive race to be the first to detect radiation and measure the spectra from a variety of high-energy objects. The Crab Nebula was an obvious target: a known source of radio and visible radiation, in 1964 Herbert Freidman from the Naval Research Laboratory showed the Crab to be a source of low-energy X-rays – the first discrete X-ray source identified – using the lunar occultation technique earlier used to identify the quasar 3C273 from Parkes. A permanent scientific balloon flight facility had been put into operation at Palestine several years earlier and universities from all parts of the US and overseas used the facility to perform scientific balloon flight experiments for a variety of atmospheric and astronomy studies. The Rice University group happened to be the only Texas institution that used the facility, and for that reason Fishman and his colleagues were told they would receive special treatment from the facility personnel. (Although in the end they didn't find that to be particularly true.)

Telescopes designed to observe the high energy universe are completely unlike optical or radio telescopes, which both have at their hearts a reflector of some description that gathers and focuses either light or radio waves to a receiver. Gamma rays cannot be focused like ordinary light, but they can at least be detected. Ordinarily such an instrument would consist of a detector in a box with a hole for the γ-rays to pass through. Although not really taking a picture, at least you could get some idea of the source of the γ-rays by noting the direction the detector was pointing in when the γ-rays were detected. When trying to observe the γ-ray sky, however, things are a little more difficult. γ-rays can penetrate the casing surrounding the 'telescope' creating a kind of haze and obscuring the object under study. For the astronomers, it was like trying to take a conventional photograph using a transparent camera. Even lead, which would ordinarily have absorbed unwanted γ-rays and X-rays, wasn't enough protection: high above the Earth's surface cosmic rays – a constant stream of high-energy particles from space – would have produced γ-rays in the lead itself making it 'glow' at γ-ray wavelengths and clouding the observation.

To get around these problems, the γ-ray 'telescope' consisted of not one but two γ-ray detectors, one on top of the other. That way, γ-rays from the object being observed had to register with both detectors in order to be recorded. Called a well detector, the alignment of the γ-ray detectors gave a line of sight to the source. The instrument was mounted on a custom-built drive system that would track the target as the balloon drifted across the sky. The entire detector system was the largest γ-ray detector ever built for balloon-flight or space-flight observations, and was certainly the most expensive and heaviest part of the telescope. The entire gondola that carried the telescope had a mass of about 600 kg. Launched just after six in the morning, the balloon reached its ceiling height three hours later and began to observe. The telescope turned as instructed and pointed towards the Crab Nebula.

All this was in June 1967. Fishman and his fellow students knew of Baade and Zwicky's suggestion that the Crab Nebula would be the resting place of a neutron star. When they asked Haymes about this, however, they were told to 'ignore' the idea. Neutron stars were, he said, an unproven notion of some wild-eyed theorists. The balloon flight was successful and the energy spectrum of the Crab was measured for the first time to an energy of several hundred keV. The team was quite proud of its accomplishment; the resulting paper put the Rice University group on the map in this field.

Then the news broke: a group of British radio astronomers had discovered several pulsating radio sources. The Rice group had no idea that pulsars were in any way connected to the high-energy observations they were performing, but when Fishman heard the news it occurred to him that there was a chance that X-ray or even gamma-ray pulses could also exist and possibly be observable from the Crab Nebula. At the time, he was busy with his thesis work involving measurements of high-energy radiation from extragalactic objects. Fishman suggested that he begin searching the data from the 1967 balloon flight for signs of a pulsing signal but was discouraged by a noted theorist at Rice. Fishman would be wasting his time, he was told, and should work on his thesis instead. 'Despite this discouragement, I decided to use our Crab balloon flight data from June 1967 to search for pulsations from the Crab,' Fishman recalls.

The X-Ray Crab

Then in late 1968, Staelin and Reifenstein announced their discovery of the Crab pulsar. Fishman was further spurred on by the announcement that the Naval Research Laboratory (NRL) X-ray rocket group had found low energy X-ray pulses from the Crab pulsar from their rocket flight observations. There were a couple of fortunate circumstances that made his search for pulses possible. Firstly, the detector was highly shielded, resulting in a low counting rate so that individual X-rays and gamma rays could be recorded as they were detected. If the X-rays and γ-rays had been binned (collected over a period of time and then recorded) there wouldn't have been enough timing resolution to detect the rapid pulses from the Crab. Secondly, at the time of the balloon-flight, the data were recorded on a multi-track tape recorder along with precise timing data from the radio station WWV, which was recorded on a parallel track. Finally, a fellow graduate student, Rick Harnden, was a computer expert and knew how to program a large computer so that the Crab observations from the 1967 balloon flight could be combined to show a weak signal from the

Crab pulsar, if it were detectable at all. The procedure was to 'fold' the data at the period of the pulsar, ~ 33 ms, over and over, throughout the recording. The problem was that the Crab hadn't been known long enough for anyone to know whether the pulse period was constant or not. For all Fishman and Harnden knew, the pulses – even if they were there – could be smeared out by folding the data at a different period than the Crab pulsar over the more than three hours of observations.

The first step was to convert the balloon flight analog tapes, recorded at Palestine and from a down-wind telemetry receiving station, into a digital format in a manner that had never been done before. Instead of summing data over discrete time intervals of source and background (normally about 10 minutes each), they had to keep the data as separate, recognizable, time-tagged photons. These data would contain the individually detected X-ray and gamma-ray photons that were time-tagged to an absolute accuracy of about a millisecond, using the WWV signal which was on the analog tape. After convincing the head of the computer facility at Rice that they had a viable research project, Fishman and Harnden were allowed to use the main campus computer for several overnight runs so that they could search many trial periods using the entire database, in order to home in on the correct one. Without this, the search would have been hopeless. For 1960s computer technology, this was a very large computational effort and required several nights of computer runs. Finally they hit the proper period that resulted in a significant pulsar signal. The pulsar's signature had been in the data all along, months before Bell's discovery, if only they had known what to look for and how.

Fishman and Harnden were anxious to announce their findings to the scientific community and to the public, as well. For the former, a 'telegram' was sent to astronomers worldwide by a communication system known as the International Astronomical Union Circulars. For the public announcement, the public relations office at Rice University hastily organized a press conference that week in the Department's large lecture hall for local reporters and correspondents from national publications. One correspondent in particular, from the World Book Science Service (no longer in operation), wrote a long feature article on the discovery that highlighted the work of the two graduate students, Fishman and Harnden, together with a photo of the pair. A large number of major newspapers in the US and overseas ran this article, as determined by the clipping service that the Rice PR office used at that time.

The young scientists were suddenly famous and were receiving calls from other scientists, friends and family who had seen the article and their photo in the local newspapers. The scientific details of the observations were written for the *Astrophysical Journal Letters* within a week and submitted. The manuscript was hand-carried by Harnden to the Editor in Chicago at the University of Chicago Press, Subramanian Chandrasekhar. He was quite impressed with the discovery and the fact that Harnden had hand-carried the manuscript from Houston to Chicago. He did not know that Harnden's father was a pilot for Braniff Airways and that Rick could fly anywhere on their system for free.

In the next chapter we will take a detailed look at the circumstances surrounding the discovery of the first optical pulsar – a pulsar that flashes on and off in visible light – which just happened to be the Crab pulsar. But before we do, it is worthwhile to examine exactly why the Crab pulsar has turned out to be of such importance. The remainder of this chapter is largely based on notes from Gerry Fishman, to whom I am grateful.

Importance of the Crab Pulsar
By Gerry Fishman

Philip Morrison once said that the Crab Nebula and its pulsar serve as a 'Rosetta Stone' for high-energy astrophysics.[9] These two objects tie together many of the ideas that we have about the formation of neutron stars and supernova remnants from a known supernova explosion in 1054 AD. Furthermore, both the nebula and the pulsar can be studied from the radio regime to the gamma-ray energies, over 18 decades in energy (or wavelength)! There have been literally thousands of observational and theoretical papers on Crab Nebula and its pulsar and there have been several books and scientific conferences and workshops devoted exclusively to these two objects and their relationships to wider fields of astrophysics. These two objects have been a favorite target of study of virtually every space-borne telescope. Combined HST and Chandra X-ray photographs and movies show spectacular images of the nebula and wisps of emitting matter emanating from the pulsar into the nebula. Prior to 1969, it was a great mystery as to how the Crab Nebula could still be giving off so much energy, almost a thousand years after its initial explosion. Of course it is now known that it is the pulsar that continually feeds energy into the Nebula and beyond by high energy particles and waves.

The Crab Nebula was one of the first radio objects discovered in the Galaxy in the late 1940s and in the X-ray region in the early 1960s, but the pulsed radio emission from the Crab was not seen until after many other radio pulsars had been discovered. This was mainly an observational problem; it was difficult to detect rapid radio pulsars. These days, pulsars with millisecond periods are routinely observed in both the radio and optical regions. At visible wavelengths, the Crab pulsar appears as a faint star within the much brighter, surrounding nebula. At X-ray wavelengths, it becomes more conspicuous and at even higher energies the pulsar completely dominates the emission from the nebula. The Crab Nebula has recently been seen with new ground-based HESS gamma-ray observatory in Africa up to TeV energies. However, the Crab pulsar is not seen at these energies and pulsed emission, if present, is less than a few percent of the flux of the Nebula.

The Crab Nebula is a dynamic object, noticeably fading over the years. Several hundred years ago, it was prominently labelled as M1 in the famous Messier catalog. In the 1950s, astronomers noted optical wisps that were moving within the nebula. But it was not until the space age, with its accompanying balloon-borne and satellite-borne instruments that the full energetics and dynamics of these two objects were observed. At X-ray and gamma-ray energies, they are one of the brightest objects in the sky. Their intensity and relative stability have allowed them to become a unit of measure at high energies. The 'Crab' and milliCrab' have become standard observational units to which other high-energy objects can be easily compared with the same instrument.

It is expected that the Crab Nebula and its pulsar will continue to play an extremely important role in astrophysics, in particular the field of high-energy astrophysics, for hundreds (or thousands) of more years, until another nearby supernova occurs in our Galaxy.

References

1. A more detailed explanation of dispersion will be given in Chapter 8: The Searchers.
2. The telescope collapsed at 9:43pm EST on Tuesday the 15th of November 1988. (See http://www.nrao.edu/archives/Timeline/300ft_after.shtml)
3. In fact even though most astronomers were confident in Baade and Minkowski's prediction of a specific star in the middle of the Crab Nebula, it would be months before the star responsible for the pulses was positively identified using optical telescopes. This is the subject of the next chapter.
4. Later studies of the Crab pulsar explained why Staelin and Reifenstein were successful. The Crab occasionally emits pulses of enormous strength separated by much weaker periods of emission lasting seconds or more, now known as the 'giant pulse' phenomenon. Because of their rare character and random appearance, these giant pulses are difficult to identify. Repetitive dispersion was their revealing signature in this case.
5. The dish consists of almost 40,000 perforated aluminum panels, each 1 meter by 2 meters, which were added not long after the Crab pulsar observations described here. The new surface improved operation of the telescope at shorter wavelengths.
6. As we have seen, the possibility of a confirmed SETI signal was a controversial and irritating factor in the delay in the announcement of the original pulsar discovery.
7. Unlike the land in the Middle East, the name of this town is pronounced 'Palesteen'.
8. γ-rays (pronounced gamma rays) and X-rays are short wavelength, high energy form of radiation, part of the same spectrum as light and the longer wavelength radio waves.
9. This was said during an interview on a TV program 'NOVA' on supernovae about twenty years ago.

7 'Optical Pulsars'

'The observational question about the optical emission also remains open; it may be settled very soon, since the whole of the astronomical world is so excited about the possibility of light pulses that many telescopes will be used on the pulsars during the next few nights.'

F.G. Smith, Pulsating Stars, May 21 1968

The three astronomers stared intently at the screen with the ambivalence of high anticipation and low expectations. Two of them were theoretical astronomers on their first real experience in an astronomical observatory. After enduring good-natured joking from more experienced observers that such novices had been allowed near a telescope, the pair had managed to book time on a rather small, almost antique telescope that was well past its initial prime. Along the way, they had rallied the support of an experienced observational astronomer who had set up much of the equipment for them. Although primarily there to gain observing experience, the theoretical astronomers had chosen as their prey one of the hottest topics in astronomy at the time: the elusive optical pulsar. More experienced observers had tried and failed to detect optical pulsars and so they searched without much hope for telltale flashes of light that kept time with the recently discovered Crab pulsar that would indicate that pulsars emitted visible light pulses as well as radio pulses. Such a discovery would be quite a prize considering its implications for understanding pulsars. In this case the sign of success would be a distinctively curved line of green dots running across the center of the monitor. That cold winter evening in 1969 the astronomers stared at the screen in darkness and anticipation.

No Optical Pulsars

At the time of the New York conference, the question remained: could pulsars be seen in visible light, or in astronomical parlance, in the optical part of the spectrum? The rapid fire radio pulses showed pulsars had to be compact objects, but radio telescopes lacked the resolution to show precisely where in the sky the pulses were coming from. If optical telescopes could zero in on the pulsars, a great deal could be learned about their true nature by studying them at wavelengths other than radio. For example, if a pulsar could be seen at optical wavelengths it would mean much more energy was being produced: it takes a lot more energy to make something glow at optical wavelengths compared with the radio. Hence, finding a visible object to match the radio pulsars was a major prize for astronomers. But identifying an optical counterpart of any of the known pulsars was not easy for the simple reason that resolution at radio wavelengths was just too poor. As Hewish and his colleagues pointed out from the beginning, the positional accuracy of the IPS Array, while permitting the discovery of CP1919, was too crude to allow any serious attempt at identifying the responsible star. 'The search area, which lies close to the galactic plane, includes two twelfth magnitude stars and a large number of weaker objects,' they pointed out in the original paper. 'In the absence of further data, only the most tentative suggestion to account for these remarkable sources can be made.'[1] Nonetheless, at least a few astronomers did make the attempt and two teams announced at the conference that they had discovered the elusive optical pulsar. What follows is an account of the events from the week of the conference to January 1969 based on the literature and the recollections of those involved in one of the most exciting stories in the history of pulsar astronomy.

Maran, Lynds and Trumbo

The scientific organizer of the New York conference, Al Cameron, had for a long time been writing theoretical papers on properties of neutron stars, including their vibration frequencies. One of his graduate students, Steve Maran, used to tease Cameron for writing so many papers on an object that was, at the time, purely theoretical. Then one day Maran was driving across the Sonoran desert from Tucson to Kitt Peak observatory when the news came on the radio that British astronomers had discovered pulsating radio sources from deep space. The British scientists further claimed these were the long sought after neutron stars. There were no mobile phones in the 1960s, and so Maran had to wait until he reached Kitt Peak Observatory before he could call Cameron to congratulate him on the discovery of neutron stars and to apologize for the teasing. 'Thanks, Steve, but no cigar yet,' Cameron replied. 'These can't be neutron stars because the vibration rates [Cameron's interpretation of the pulse rates] are too slow.' A day later Cameron called Maran at Kitt Peak and told him he thought the new pulsars were in fact white dwarfs that were vibrating faster than expected. This would require modest modifications to white dwarf theory, but Maran should tell the leading observer at Kitt Peak, Roger Lynds, to look for optical pulses. Maran walked down the hall to Lynds' office and conveyed the message. Within an hour Lynds telephoned Maran and said he knew how to search for the optical pulses. The problem was, he had no immediate access to any of the telescopes. It just so happened that Maran had what was effectively a telescope all to himself. The 1.3 meter telescope on Kitt Peak was at the time an experimental, remote control telescope to which Maran had sole access. Lynds suggested that if Maran could provide the telescope, he would provide the instrumentation and together they could search for optical pulses from white dwarfs. With the help of the head of electronics at Kitt Peak, Donald Trumbo, Lynds put together the necessary equipment.

The astronomers turned the 1.3 meter telescope to CP 1919 looking for any sign of optical pulses. In order to detect optical pulses, it was necessary to collect the light from the pulsar and measure its intensity at a fraction of the radio pulse period. The way to do this was to attach a device called a photometer that converts light gathered by a telescope to an electrical signal. The intensity of the signal is then measured in order to detect any rise and fall of light intensity that would indicate optical flashes in time with the radio pulses. At first they seemed to have been successful and announced the discovery at the New York meeting. The variation in brightness was tiny – just 4% of the average brightness of the star – and it varied at twice the rate of the radio pulsations which was surely no coincidence. The idea was puzzling for the theorists, since the question now was what could make the optical pulses appear on every alternate rotation of the pulsar? The following day a bleary-eyed astronomer, David Cudaback, arrived from Lick Observatory where he had been observing CP 1919. Cudaback not only 'confirmed' the optical pulses, but reported an unexpected variation in the pulse period, which was speeding up and slowing down by as much as 10%. Optical pulses seemed to have been found, albeit in an unexpected form. Unfortunately, observations made from other observatories over the following months failed to confirm either set of results. The claims to have detected optical pulses from CP 1919 was premature, although to this day the reason why remains unclear. Maran and Lynds attempted searches of the other three pulsars announced at the time but without success. Eventually, Maran and Lynds agreed to cease their search for optical pulses, and the equipment they had put together for their search sat idle. Optical pulses from pulsars remained invisible.

Willstrop

Within six weeks of the publication of Hewish and Bell's discovery, astronomers at Cambridge were also searching for optical pulses from CP 1919 that kept time with the radio pulses. Using the 0.9 meter telescope at the Institute of Astronomy, Roderick Willstrop and John Jelley carried out such observations in the early hours of April 5th and 6th 1968, but no evidence of optical flashes from CP 1919 was found.

The equipment they used could only be set to search for one pulse frequency at a time, and so when pulsars with different pulse rates were discovered, it became clear a more versatile method of recording and an improved photometer were both needed. Willstrop set about designing a photometer in which a wheel with four filters could be placed in the beam of the telescope allowing him to more quickly and accurately measure the 'color' of the objects being studied. Meanwhile Ed Kibblewhite designed and built a photometric recording system which used a high-speed punch for recording the output of the photometer. The latter could perforate seven track paper tape at up to 110 rows per second. The paper tape was allowed to fall, or perhaps the word 'spew' was more appropriate when the punch was run near to its maximum rate, into a wooden box (a tea chest, about 60 cm cube). At the end of each observation the tape was wound up using an adaptation of a hand-powered grinding wheel and later taken to the Cambridge University Computer Laboratory and analyzed for any periodicities.

There were two possibilities for the sources of the radio pulsations: white dwarfs or neutron stars. White dwarfs weren't the most likely candidates for pulsars, but they hadn't yet been ruled out when Vaughan and Large's paper in *Nature* announced the discovery of the Vela pulsar. This was followed by the December issue of *Science* reporting the discovery by Staelin and Reifenstein of the Crab pulsar. As we saw in the last chapter, these discoveries were not coincidental: two high speed pulsars associated with supernova remnants was dramatic confirmation of Baade and Zwicky's prediction made some 30 years earlier. Pulsars had to be neutron stars, not white dwarfs. Nonetheless, Willstrop already had time booked on the 0.9 meter telescope and he was planning to use the new photometer to look at white dwarf stars to eliminate them as possible sources of optical pulses. On November 23rd 1968 the sky was clouded and Willstrop went to bed at 9:30pm. But the sky cleared an hour later and he went out to observe. Willstrop's notebook records:

> At 23.36 decided to start on Crab Neb. Found it with no difficulty. Can see faint starlike condensations. If I can find a photo I might identify the central object. Went to library at 23.55. Cloudy at 00.10 !! Carried out various tests. Reopened at 00.36 Clear patch. M1. 1 mm diaphragm, no guiding. Start at 00.45 Stop after 50 seconds. We are looking through cloud. Raining at 01.02 Cleared at 02.15. Reopened, set on AGK2 +21 541, then offset 2 minutes to Crab. Noted the star field, and just identified these stars on Sky and Telescope vol. 28, p. 334, 1964 by 02.30. 02.36 Saturn? visible in W. hope !
>
> M1 02.56.30 > 11.5 mins
> 96/second Prescaler 4 Typical reading 15 or 16
> 1 mm diaphragm – no guiding.

Willstrop made further observations that night, and again in the early hours of the 25th, and over the next week spent more time observing white dwarfs. The accumulated paper tapes containing the data from the white dwarf observations and the supposed neutron star in the center of the Crab Nebula were taken one or two at a time to the Computer Laboratory for analysis. It was December 1968.

Cocke and Disney

In August that year two young theoretical astronomers from either side of the Atlantic arrived in Tucson, Arizona, each unaware of their mutual destination: Steward Observatory. This observatory is run by the University of Arizona at

Tucson, whose main telescopes are also located on Kitt Peak. John Cocke was an American astronomer from NASA Goddard Institute for Space Studies; Michael Disney was an outgoing British astronomer from the University of London Observatory, Mill Hill. By sheer coincidence, they met at the pool side of a motel they and their wives had booked into as a temporary residence while they looked for somewhere permanent to stay. They began talking and realized they were both heading for the same observatory. Neither had any experience with using telescopes; Disney in particular felt out of his depth as there were so many experienced telescope users while he was a theoretical astronomer. 'I didn't really know what a telescope looked like,' he later admitted. With their inexperience in mind, the two astronomers decided it would be a good idea to remedy the situation by actually booking some telescope time and having a go.

Cocke and Disney applied for telescope time and, to their surprise, were granted a few nights on a relatively small 90 cm telescope; the managers of the observatory felt the pair would only waste a few nights, so it would be no great loss to the observatory. The telescope was one of those instruments rich in history but well past its prime. It has been said that telescopes have a useful life of about 50 years; they don't necessarily wear out, but as technology advances they are surpassed in size and sensitivity. By that standard the 0.9 meter telescope, built in 1922, was elderly when Cocke and Disney entered the dome-shaped enclosure for the first time. The telescope was originally erected on the campus of the University of Arizona in Tucson, and was the first built of entirely American-made products. As the city grew, so the skies brightened and in 1962 the aging telescope was moved to the darker skies of Kitt Peak. The telescope is now dwarfed by much larger instruments, but Kitt Peak Observatory owes its existence at least in part to the 0.9 meter telescope. The mountain is part of an Indian reservation and considered sacred by the Indians. While it was still in Tucson the 0.9 meter telescope had helped to win the acceptance of telescopes on Kitt Peak by giving the tribal council a view of the heavens. By the time Cocke and Disney arrived, the telescope was regarded as small and ancient.

Cocke and Disney now had their telescope time, but what to look at? Although they knew their chances of success were slim, they decided to look for pulsars in the optical. Cocke asked a number of prominent astrophysicists about the possibility of detecting a pulsar at visible wavelengths, but they were all very negative about it and told him he'd be wasting his time. Since it takes much more energy for an object to shine in visible light than it does at radio wavelengths, it was generally agreed that pulsars would be too distant and faint to be detectable at optical wavelengths. Not only that, but other, more experienced observers had been searching for optical pulses since May without success. Nonetheless, the purpose was to gain telescope experience and so Cocke and Disney felt they had nothing to lose. But where to look? The resolution of radio telescopes being far less than that of optical telescopes, the coordinates of the discovered pulsars weren't exactly precise. The Vela and Crab pulsar discoveries changed all that by associating a radio object with an optical counterpart, a supernova remnant. While the Vela supernova remnant is a spread out, wispy filament stretching across the sky, the Crab is much more compact, being about a fifth the diameter of the full Moon.[2] As we saw in the last chapter, Lovelace and his colleagues managed to pin down the position of the Crab pulsar to within 5 minutes of arc from the center of the nebula itself. If the pulsar truly was the remains of the star that led to the Crab supernova, then the logical place to start looking would be in the middle of the Crab Nebula itself. There are not one but two stars in the center of the Crab that stand out from the others, one of which was identified by Baade and Minkowski in 1942 as the likely remains of the supernova, and appropriately named Baade's star. While there are other stars seen 'within' the nebula,[3] Cocke and Disney decided to see if Baade's star was pulsing in the optical at the same rate as the radio source.

The Search Begins

Searching for an optical pulsar isn't as simple as looking through a telescope or even using electronic equipment to send an image to a monitor and watch for a flashing light. If the suspected pulsar was indeed flashing on and off at the same rate as the radio pulses, which in the case of the Crab pulsar is every 0.033 seconds, the slowly responding human eye would see nothing more than a steady glow. What was needed was an alternative method of detecting the pulses, which meant equipment and expertize that neither Cocke nor Disney had. One of the astronomers at Steward Observatory, Ray Weymann, suggested they ask another astronomer who also worked at the observatory. That astronomer was Don Taylor.

Taylor is an electronics wiz who at the time was involved in a project to see if quasars had underlying galaxies. To do this he used a technique known as 'area scanning'. This involved mounting a device with a small slit on the telescope just where the image was produced. By moving the slit back and forth across the field of view, the amount of light passing through the slit would vary as it passed across the object being observed. Each time the slit passed across a quasar, for example, the amount of light passing through the slit rose and fell producing a light curve, a sort of graph of the intensity of the light compared with position across the field of view. But quasars, at least seen from Earth, are faint and this meant Taylor had to look at a given quasar for a long time in order to build up enough light to study it properly. Herein lies a problem that led him to develop equipment that turned out to be ideal for searching for pulsars. Because the Earth's atmosphere is continually shifting and distorting the light of celestial objects, a single slow scan tends to be noisy and distorted. A quicker scan tends to be more accurate, like taking a short exposure photograph of a moving object, but it also means less light makes it through the aperture resulting in a fainter signal. By combining the signals from many brief scans and taking an average a more distinct light curve can be built up. This averaging was done in a special type of computer called a CAT which stands for Computer of Average Transients. The CAT took each scan and combined them to produce an average curve that represented the way the brightness of the quasar varied from one side to the other.

It turned out this was just the sort of thing that was needed by Cocke and Disney. The anticipated optical pulses from a pulsar would also be a series of repeating but faint and noisy light curves: each time the pulsar flashed on and off its brightness would rise and fall producing a light curve, with each curve representing an individual optical pulse. If the telescope was indeed collecting the light from a pulsar, the amount of light collected by the telescope would be rising and falling at the same rate the pulsar was flashing on and off at radio wavelengths. The CAT could sample the light at the desired frequency then combine the individual light curves from many individual pulses, displaying them as a single curve on a screen. But how to get the CAT to sample at the required pulse period? In area scanning the CAT is told when each scan is about to start at the telescope and so it knows where each light curve begins. With the optical pulsar search, the CAT would need another way of knowing when to start each curve, and that required a device called a frequency synthesizer. This device could be set at the same period as the pulsar under observation as determined by the radio astronomers. Its job would be to tell the CAT at what rate to sample the data, in other words when each light curve was about to begin. Although Taylor did not have one, it just so happened that on the same mountain such a piece of equipment lay unused following an earlier attempt to find the elusive optical pulsar.

Maran and Lynds Loan Equipment

John Cocke had never met Maran or Lynds but made the short journey from his office at Steward Observatory across Cherry Avenue to the headquarters of the Kitt Peak National Observatory. He explained to Maran that he was a theoretical astronomer, but that he had a hunch that the fastest pulsars would produce the brightest pulses. Cocke went on to say that he and Disney wanted to look for optical pulses from the recently discovered – and certainly fast – Crab pulsar. Could Maran provide any equipment? Now, Maran and Lynds had already shown that, whatever was producing optical pulses, they were fainter than just about anything astronomers had detected at the time. Not only had they used a 1.3 meter telescope in their efforts, they had been able to integrate their observations over several nights allowing them to reduce the background lighting electronically, in effect increasing the contrast between the pulsar signal and the background sky. All of their efforts had come to nothing. With this experience in mind, Maran and Lynds thought searching for optical pulses from the Crab was silly but nonetheless loaned Cocke their frequency synthesizer and another device, a pulse amplifier. This device detects individual photons of light gathered by the telescope it is attached to and converts them to electrical signals that can be recorded and analyzed by the CAT.

There was one more piece of equipment they needed: a photometer. Cocke, Disney and Taylor borrowed one from another astronomer also on the faculty at Steward Observatory, Walt Fitch. The Fitch photometer had been built specifically for use on the 0.9 meter telescope and had been used by Fitch as well as other astronomers at Steward. Cocke, Disney and Taylor now had the equipment and the telescope. It was time to gain some experience.

The Search Begins

Friday, January, 10th 1969. The 0.9 meter telescope was of a Newtonian design, common in amateur telescopes but not used for professional telescopes these days. When the telescope it pointed at the night sky, it's business end – where the observer or detector sits – is way up in the air and so to attach equipment the astronomers had to climb onto a gantry high above the floor of the observatory dome. The dome itself was kept in total darkness to allow the telescope and its detector to sense the faint signals from the pulsar. To make matters worse, Kitt Peak in the middle of winter is bitterly cold and Disney, who hadn't anticipated this, had worn only a sports jacket. Risking a serious fall in the dark (the gantry had no safety railing) combined with the cold, it was certainly a hard introduction to observational astronomy. Disney began wondering why on Earth he was there at all. Taylor and a night assistant,[4] Bob McCallister, set up and then monitored the electronics while Cocke and Disney gained experience using the telescope itself, learning how to locate the pulsar suspects. In the dark, in the cold, they saw nothing. Then the clouds rolled in and ended Cocke and Disney's first night using a telescope.

Cocke, Disney and Taylor returned to the telescope the following night and made their first serious detection attempt. Although the equipment and the telescope seemed to be operating perfectly, again there was nothing. The third night of their precious observing time passed without incident and without a result. Then the weather closed in: their last two nights at the telescope were ruined and they had seen nothing. Then luck stepped in. The astronomer who was next in line to use the telescope, Bill Tifft, called to say his wife had become ill and so he was unable to use the telescope. Would Cocke and Disney like the extra two nights observing time? Naturally they accepted the offer.

Taylor had returned to the University campus in Tucson: his students were approaching exams and he wanted to make himself available to them. He reasoned that now that the electronics had been set up and were running fine there was no immediate need for him to be there. Besides, they clearly needed a larger telescope and had already booked time on the Lunar and Planetary Laboratory's 1.6 meter telescope[5] for the nights of the 15th and 16th February. Taylor saw no great urgency in being around for what were likely to be more negative results using the 0.9 meter telescope. McCallister remained to help the two astronomers.

FACULTY & STAFF Newsletter

THE UNIVERSITY OF ARIZONA   February 1969

ix times

me how I
Councilman
On the sur-
. Work on
mity to en-
· hand, aca-
erve better

ortant cave-
., the temp-
issions into
anecdotes
political ex-
vant to the
t the case,
the academ-
not always
k of urban
ouncil. In
ment have
o solve the
community.
e that serv-
cial is not
zen politics
or most of
olding roles
reference,
ng regular,
life is very

e enormous
campaign.
e fact that
ts formally
olvement in

UA astronomers who made the first optical telescope sighting of a pulsar are, from left, Drs. Michael J. Disney, W. John Cocke and Donald J. Taylor. They are shown with electronic devices used in the discovery.   (Bob Broder photo)

Pulsar Sighting By UA Astronomers Hailed As Unusual Scientific Feat

Three University of Arizona astronomers are tired but happy with their optical dis-

first picked up with radio telescopes in 1967. The 30-times-per-second pulse rate

Figure 7 Mike Disney, John Cocke and Don Taylor at the time of the discovery of the optical pulses from the Crab pulsar. (Courtesy John Cocke and Don Taylor.)

Time to Rethink

While waiting for the sky to clear, Cocke decided he should recheck his calculations. They were discussing the project over dinner when Cocke pointed out that as Earth orbits the Sun it is sometimes moving towards the pulsar and sometimes away from it. It's a bit like looking at a friend standing to the side of a merry-go-round as you ride past them. The Doppler effect meant the pulsar's rate would increase as Earth moved towards the pulsar, and reduce it as it moved away. It was then they realized they'd made a mistake with their calculation of the pulsar period: they'd been looking at the wrong period. Cocke felt foolish at having miscalculated the correction for Earth's motion. Despite this, and the feeling of futility of the whole project, he went over all the calculations again. That night the three of them – Cocke, Disney and McCallister – went to the dome and reset the timing equipment.

The Sixth Night: Thursday 15th January 1969

Once the telescope had been set it more or less ran itself. As before, the astronomers pointed the 0.9 meter telescope at the heart of the Crab Nebula. The three crouched down in front of the monitor. Almost instantly a wave of green dots appeared right in the middle of the monitor. The three of them stared incredulously at the monitor. Disney finally exclaimed : 'We've got a bleeding pulse here!'. The monitor showed a wave that could only mean the telescope was indeed picking up optical flashes in time with the radio pulsar in the Crab Nebula. Cocke said, 'I hope it's an historic moment. We'll know when we take another reading. And if that spike there is again right in the middle…that spike is right in the middle and that scares me.' With a mixture of excitement, doubt and incredulity they began to check any obvious sources of error. The equipment was turned off and then on again to make sure it wasn't an artifact of the electronics. The pulses returned. 'My God, it's still there,' Disney said. 'It's as good as it was, or better than it was last time.' The three were almost hysterical with excitement and it took them a few moments to regain their composure. Eventually, Cocke suggested they point the telescope away from the Crab to see if the signal disappeared. It did, but not completely. A faint pulse remained even though the telescope wasn't pointing directly at either star. So they changed the frequency that the computer was running at to confirm that the source of the pulses wasn't in the equipment. When they did the signal disappeared; the pulses they were seeing were definitely not coming from the electronics. Disney then suggested that perhaps the telescope wasn't moved far enough away from the nebula, that a small amount of light from the flashing star was 'leaking' into the telescope. To eliminate this possibility, they moved the telescope so that it was pointing to an entirely different part of the Crab Nebula. The pulses disappeared entirely. They returned to the Crab, and there was the pulsar flickering on and off. It was now without doubt: Cocke and Disney had detected the first optical pulsar.

Disney pointed out that it was a terrible shame that Don Taylor wasn't there to participate in the actual discovery, especially since he had done so much towards making it possible. Disney phoned him: 'Don, we found the pulsar'. He went on to explain to Taylor about the timing error and how they had corrected for it, but Taylor remained skeptical. 'The false detections made me all the more cautious to not make a similar mistake,' he recalled to me. Taylor made several suggestions as to what the pulses might be other than the actual pulsar. Cocke and Disney were told to conduct some standard tests, including pointing the telescope away from the pulsar and changing the CAT sweep period to see if the pulse remained. There was nothing obviously wrong with the observation, and yet he told Disney to keep the discovery quiet until he had a chance to conduct a further run with the telescope to confirm the discovery. Taylor's concerns were not only about some technical error, but also about other astronomers with access to larger telescopes taking over the discovery. He was prepared to drive up to the observatory that night, but Disney said not to bother since cloud was yet again rolling in. After Disney had hung up the phone, the three at the observatory spent the rest of the night repeating the observation using different colored filters, and making other variations to the experiment.

The Seventh Night: Friday 16th January 1969

The following evening, the last night at the telescope, it was clear. They set up the observation again for Taylor, who repeated a number of tests to make sure they weren't being fooled by the electronics. It wasn't long before Taylor was as convinced as his colleagues: optical pulsations from a pulsar had indeed been found. But there was still an unanswered question: which star was the pulsar? As I mentioned earlier, there are not one but two stars that appear to be in the center of the Crab Nebula, either one of which could have been the source of the pulsations. When it comes to distinguishing between stars close together, astronomers refer to which star is north or south, and preceding or following the other. The latter part of the description comes from the fact that as you look through a telescope that is stationary, objects seem to move from the east to the west across the field of view like members of a parade. The apparent east-west movement is caused by the telescope being carried by Earth's rotation. Looking through the telescope, objects to the west precede those to the east, which trail behind, or follow. The two stars in the middle of the Crab Nebula were hence referred to as 'south preceding' and 'north following'. Although the two stars are of similar apparent brightness, in 1942 Walter Baade ruled out the north following star as the source of energy for the Crab Nebula. Ever since the north following star was not generally regarded as being physically associated with the Crab. During the observing sessions, the observations were recorded on audio tape for convenience. The tape was left running accidentally and this is how we now have a candid record of the unfolding events. On one of the tapes, the astronomers clearly state their opinion that the pulsar they had detected was the south preceding star. Despite the fact that no one at the time seriously thought the north following star was the pulsar, Cocke, Disney and Taylor had to be sure.

To positively identify the optical pulsar, the astronomers had to single out the light from first one star and then the other to see which one was emitting the pulses. The difficulty for Cocke, Disney, and Taylor was that the two stars were so close together compared with the field of view of their telescope. In order to look at one star at a time, they needed a diaphragm – a thin metal plate with a tiny hole drilled in it – that was small enough to block out all but the star being observed. The diaphragms used on the 0.9 meter telescope up to that point were too big and allowed the light from both stars into the photometer simultaneously. To give you an idea of what they were up against, you need to understand how tiny a distance on the sky they were dealing with. The distances between celestial objects is given in terms of the angle they make with the observer and are given in degrees, minutes (one degree equals 60 minutes, or 60′) and seconds (1′ equals 60 seconds, or 60″). The Moon, for example, is about half a degree across, while the star alpha Centauri is actually two stars separated by just under 18″, and can easily be seen as two stars in a small telescope. But the separation of the double star in the Crab is much smaller, a mere 4″, and so in order to let through the light of one star at a time calls for a diaphragm of similar or smaller diameter.

The diaphragm Cocke, Disney and Taylor were using was 22″ across and far too large to positively identify the pulsar so they tried an alternative technique. After they had confirmed the optical pulses for Taylor, the three set about trying to narrow down the most likely position of the source of the pulsations by carefully moving the telescope away from the center of the Crab Nebula and watching for how far the telescope could be moved before the pulsations disappeared. Using this astronomical version of groping in the dark they were able to suggest in their IAU circular:

> The estimated position of the pulsar is 5″ north and 4″ east of the south preceding star of the Crab central double. The error in position is estimated at ±5″.

This placed the likely source of the optical pulsations frustratingly close to the position of the north following star but also including the south preceding star. Nobody believed the north following star was really the pulsar, but they were obliged by the rules of science to report what they had seen, not what they suspected. In the subsequent paper published in the journal *Nature* the three wrote they 'tentatively' identified the pulsar as the south preceding star. What they really needed was a larger telescope and/or a smaller diaphragm to single out each star in turn and thereby prove which was the actual pulsar.

At this point in the story there is some conflict between the recollections of those involved. For example, while it is agreed there was some discussion between Cocke and Lynds about whether the three new astronomers were able to use a much larger 2.1 meter telescope with Maran and Lynds to search for the actual star that was producing the pulses, it is unclear after so many years who made and who declined the suggestion. At any rate the idea fell through. What was clear was that the word of the discovery was now out, and it suddenly became a race to identify the source of the optical pulses. Convinced they could identify which of the two stars in the Crab was the pulsar, Maran and Lynds wasted no time in pursuing their prey. The 2.1 meter telescope was already capable of producing larger images than the 0.9 meter telescope Cocke, Disney and Taylor were using, and it was equipped with diaphragms down to 1″ diameter. Since they thought it would be unreasonable to ask for their equipment back considering the circumstances, they cobbled together the necessary electronics and prepared the 2.1 meter telescope for observing that night. Now there were two teams on the mountain looking for the definitive source of the optical pulsations.

Limited to their smaller telescope,[6] Cocke, Disney, Taylor quickly needed a smaller diaphragm. With Bob McAllister, they set about making a smaller diaphragm for their telescope. Using tweezers and razor blades and peering down a microscope found somewhere in the building, they tried desperately to punch a hole a tenth of a millimeter across in some aluminium foil from a TV dinner. Eventually, about five in the afternoon, they succeeded. Cocke and Disney mounted their tiny diaphragm into the telescope, and as soon as it was dark they began observing. This time, however, rather than watching a monitor, Taylor hit on the idea of listening to the photometer. When the telescope was perfectly centered on the actual pulsar, Taylor would hear the pulsations as a kind of rapid clicking like a playing card held against the spokes of a spinning bicycle wheel. Shortly before nine in the evening Taylor heard the tell tale clicking showing that the telescope was pointing at the pulsar.

But according to Lynds, they were too late. While they were observing the phone rang at the dome of the 0.9 meter telescope. Taylor answered the phone: it was Beverly Lynds, Roger Lynds wife, also an astronomer and Assistant Director of the observatory. Beverly said 'You can quit [trying to identify the pulsar now] now because Roger has identified it as the south preceding star.' Maran and Lynds had used the larger telescope to distinguish between the two stars. By centering on one star at a time and completely eliminating the light from the other, within minutes they identified the star responsible for the optical pulsations: Baade's star.

Importance of the Optical Crab Pulsar

The discovery that the Crab pulsar is emitting visible light is important for several reasons. Firstly, it's a matter of energy output. It took an enormous radio telescope to detect the Crab's radio pulses, and yet a relatively small optical telescope was used to detect the visible light pulses. The implication is that much more energy is being emitted as visible light than in the radio. Further, the visibility of the Crab Nebula 900 years after the event was a bit of a mystery. As Gerry Fishman expressed it to me, it's a bit like hearing a fire cracker the day after it exploded. We now know that the nebula is being continually fed by the pulsar. As we saw towards the end of the previous chapter, the Crab has now become one of the most important pulsars studied, and it is likely to remain so.

From this point on, pulsar astronomy was to establish itself as a new field of science. Both observational and theoretical work bloomed in the wake of the extraordinary discoveries of 1967/68. The remainder of this book explores the amazing evolution of pulsar astronomy and how it has come to impact a surprising range of astronomical endeavours: from the discovery of the first planets beyond our Solar System to the confirmation of general relativity.

Epilogue: Prediscovery Detection . . . Again

On Saturday, 18th January, news of the detection of optical pulses from the Crab Nebula reached Willstrop. Until then there had been a number of uncertainties in Willstrop's mind about the likelihood that anything significant had been achieved during his searches for optical pulses. For example, although he had observed the Crab, he had concentrated on searching white dwarfs which were one of the most promising candidates for pulsars. 'I had accumulated a stack of data on white dwarf stars,' Willstrop told me. 'Hewish et al. had postulated that pulsations of white dwarf stars were the origin of the radio pulses first detected from CP1919 etc., and one of my aims was to show whether or not white dwarf stars showed fast optical pulsations.' Besides, there were serious doubts about how well the Crab observations had gone. The accuracy of the radio position was uncertain by up to 10 arc minutes potentially placing it well outside the Crab Nebula. Not only that, even though he used a 1mm diaphragm with his photometer it still gave a field of view of 50 arc seconds, enough to allow a lot of the background sky light at Cambridge into the photometer drowning the pulsar signal. The uncertainty in position of the pulsar meant there was an impractically large area of sky to search for something that may not exist, and so he simply pointed the telescope as close as he could judge to the center of the Crab Nebula (Baade's star was too faint to see except on the most transparent nights) and let the photometer run. With these doubts and the slowness of the computing process in 1968, Willstrop decided to concentrate on the data collected from white dwarfs.

During the Sunday afternoon he carried out an order-of-magnitude calculation and estimated that the observation he had already made in November might reveal the flashes, and towards midnight he observed the Crab again. The IAU Circular formally announcing Cocke, Disney and Taylor's discovery arrived at Cambridge on 24th January. Willstrop analyzed the observations of the Crab he had made two months earlier: there all along were the optical signals from the Crab pulsar. Had Willstrop examined the smaller data set from the Crab first, he would have discovered the first optical pulsar.[7]

References

1. A. Hewish *et al. Nature* 217, February 24, 1968.
2. The Crab Nebula is roughly oval shaped and measures about 6' by 4'. It is visible in amateur telescopes under dark skies.
3. As it turned out, all the stars other than Baade's star are background stars not associated with the Crab.
4. A night assistant is a permanent observatory staff member whose job it is to run the telescopes for the visiting astronomers.
5. This telescope was also being used to map the Moon's surface in support of the Apollo missions to the Moon.
6. As an aside, within a decade of those historic nights, the little telescope that had revolutionized pulsar astronomy fell into disuse. In 1982, however, it was handed over to the Spacewatch project, a search for near-Earth asteroids that pass perilously close to Earth and risk impact. It was fully refurbished and began observations in 1983. It continues to be an integral part of the Spacewatch program.
7. Roderick Willstrop's entire account of the episode is reproduced in the Appendix on pp. 181–186.

8 'The Searchers'

Why Search for Pulsars?

Having established the existence of pulsars, the next natural questions were: how many are out there, and are they all the same? There are several reasons why astronomers want to know. For one thing, since pulsars are clearly an important aspect of the story of stellar evolution, astronomers wanted to know just how many supernovae spawn pulsars and how often? It was also known that the pulsars radiate away angular momentum causing them to slow down as they age. It followed that if you could determine the slow-down rate of a pulsar, it would be one indication of how old it is. There were other factors to be explored about pulsars, too, such as how much energy they gave off, and how stable the periods were over time. Also, pulsars are made of some pretty strange stuff so represented a way of exploring the behavior of high density matter. But the observation of pulsars isn't confined to pulsars themselves; they also act as a probe of the interstellar medium. The radiation from a pulsar is subtly changed during its journey from the pulsar and Earth. Because the beam is so narrow and precise it passes through a long, thin sample of the intervening space. The features of this sample are imprinted on the beam when it arrives at Earth. For example, magnetism affects the way the beam is polarized and so by studying the polarization of pulsar beams in many directions astronomers can build up a three dimensional map of the magnetic Milky Way.

These was reason enough to undertake a major search for pulsars in the Galaxy, but perhaps the most important justification of pulsar surveys – which was unexpected in the early 1970s – was the discovery of new types of pulsars. Over the next few chapters we will look at some examples of these discoveries, but before we do it is worthwhile looking at how the early pulsar surveys were carried out. This has been an international effort (see box), and it is impractical and a little tedious to describe in detail each of the surveys. However, here is a brief description of how astronomers began searching for pulsars, a search that revealed a Galaxy sparkling with tiny radio stars.

Factors Affecting the Search for Pulsars

The first pulsars were an unexpected find; no one was deliberately searching for them. Despite the prediction of neutron stars decades earlier, astronomers held out little hope of ever discovering such objects. It was only through their peculiar

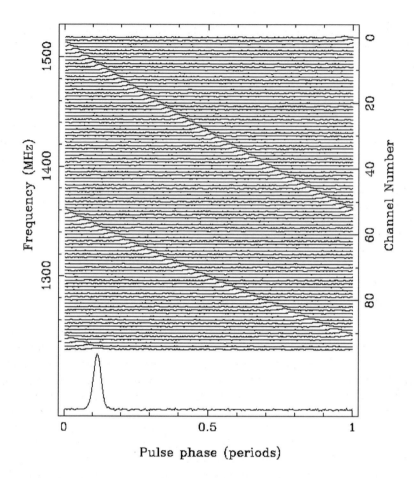

Figure 8a Plot showing how each pulse from a pulsar is dispersed across a range of frequencies. The signals from pulsars need to be 'de-dispersed' before they can be analyzed. From *Handbook of Pulsar Astronomy* (Lorimer & Kramer 2005). (Courtesy Duncan Lorimer.)

radio emissions that they were discovered at all: they were point radio sources and they flicked on and off with startling rapidity and regularity. In addition, the original pulsar discoveries were themselves fortuitous in that, using modern survey techniques, there is the chance they also would have been overlooked. Take the Crab pulsar, for example. Although it is a very bright radio source, it was only the Crab's 'giant pulses' that drew attention to itself as a pulsar. Even the Vela pulsar called for some ingenious techniques in order to be identified as a pulsar. However, remember all of this occurred only after the original pulsars had been found by Hewish and Bell, and so astronomers had at least a rough idea of what to look for.

There are other factors that make most pulsars difficult to find. For one thing, most pulsars aren't terribly bright. Although intensely hot and energetic, they are very small compared with the stars you see in the night sky. And like any small object viewed from a distance, they are hard to see. In fact the majority of pulsars are too far away to be detected even with modern search techniques, and many are on the other side of the Galaxy and so forever hidden behind a veil of gas and dust. Only the strongest pulsars, including CP 1919 and the Crab and Vela pulsars, were detectable as a series of individual pulses. Pulsars don't live forever, either. With time, their beams lose their intensity and so the pulsar gradually fades from view. Of course, all this happens over galactic time of millions of years, but when astronomers look for pulsars they are taking a snap shot of the pulsars visible in this millennium, a tiny fraction of the lifetime of even the massive stars that spawn pulsars. Even then, of the pulsars that are shining at the moment, a pulsar can only be seen if its beam happens to pass across Earth and as long as the beam is being emitted. Given these and other difficulties, just how do astronomers look for pulsars?

Flash Photography

When astronomers first began seriously searching for pulsars, they needed to determine what they were looking for and then design a search sensitive to those characteristics. Of those pulsars still bright enough to be seen, in an accessible region of the Galaxy and at the correct orientation so that their beams cross Earth, there was one other major factor that pulsar astronomers had to come to terms with: the brevity and incredible repetitive frequency of the radio pulses themselves. Without a method of analysing a radio source in fine slices of time, the signal from a pulsar blurs into a constant hum of radio emission. This was the same problem that Cocke, Disney and Taylor had to overcome when searching for the optical pulsar. Hewish's Interplanetary Scintillation (IPS) array was able to detect CP 1919 because the telescope was so sensitive in time. But CP 1919 is a relatively slow pulsar; to find faster ones called for an entirely new approach. It was the equivalent of the difference between taking a portrait photo and a sports photo. In both cases the exposure is important, but with portrait photos the photographer can concentrate less on the exposure time. Sports photography calls for much shorter exposures in order to distinguish the instant action of the subject. In a similar way, radio astronomers had been used to a virtually static radio Galaxy: although it moved, it did so with such lethargy that exposure times were not an issue. Suddenly there were these radio sources that changed with such rapidity they called for a new technique to distinguish them from the background radio sky.

To deal with this, a new technique was developed that relied heavily on computers. First, the signal is recorded on magnetic tape. The recording is then analyzed by sampling the radio signal to see if its intensity changed periodically. A pulsar's signal doesn't rise and fall like a sine wave, however. Rather it is flat for most of the time and then has a little spike representing the passing of the beam across Earth. For this reason, each signal has to be analyzed at least ten times faster than the pulsar period or it would likely be missed. To make matters more difficult, the period, dispersion and pulse width of each pulsar is unique and, until discovered, completely unknown. Astronomers have to sample the data within a range of periods and dispersions to maximize their chances of detecting a pulsar.

Once the period of the pulsar has been determined, the individual pulse profile can be determined. This is done by adding together several hundred pulses so that they all overlap in time, creating an average pulse profile. Such a profile has three main characteristics: the period (how long between one pulse and the next), the width of the pulse itself, and dispersion measure (which we looked at in Chapter 6). The pulse profile can range from single peaks to complicated double or multiple peaks, each profile reflecting the individual magnetic structure and orientation of the neutron star. While the periodicity of pulsars – the time between one pulse and the next – is extraordinarily accurate, the individual pulses can vary quite a bit. The important thing here is that each pulse profile is as unique as your finger prints.

Dispersion

So far, so good. But there remains a problem. In Chapter 6 we discussed how Staelin, Reifenstein, Brundage and Braly developed the necessary hardware and software to deal with dispersion of the radio pulses as they crossed interstellar space. It is now time to take a more detailed look at this phenomenon. Most natural sources of radiation – whether they be light, radio, X-rays or whatever – shine over a range of wavelengths; they don't emit radiation in a single wavelength or color. Pulsars are no different. Each radio pulse from a pulsar is emitted over a range – or band – of radio wavelengths. You'll recall that dispersion is the delay in the arrival time of each pulse from a pulsar when seen at different wavelengths: a pulse seen at longer wavelengths arrives at Earth behind the same pulse emitted at shorter wavelengths. Such an important aspect of pulsar searching, calls for an analogy to make sure it is properly understood.

Imagine we were setting up a race for a line of ten equally fast sprinters. Not only that, let's dress them all the same, and otherwise make them indistinguishable. The original track is a smooth, dry, even surface with the lanes clearly marked. For sprinter number one, we'll make no modifications to the track: he has the smooth, dry track with lots of traction. A fair run for him. To make things more interesting for runner 2, we spray water along his lane to create a wet and slippery surface. Let's make the track for the third runner even more difficult by sprinkling pebbles randomly along the length of the track. For runner 4 we'll add a thin layer of gravel. Let's add stepping stones for runner 5's track. For each runner down the line we'll make the surface increasingly difficult. Poor old runner number 10 has to battle his way over an obstacle course of large boulders to get to the finish line. This should be an interesting race! The gun fires and the race begins; each runner is off at an even start. As the race proceeds, however, runner 1 on the smooth track pulls ahead of runner 2, who is in turn pulling ahead of runner 3 who is occasionally slipping on the randomly sprinkled pebbles. Runner 4 is having a terrible time gaining traction on the gravel path. Each runner has an increasingly difficult time running to the finish line. Poor old runner 10 is making slow progress, and is now falling well behind runner 1. At the end of the race, the arrival times of the runners will be later for the runners with more difficult surfaces. An observer at one end of the finish line would see a constant line of competing runners, a trail of people all exactly the same, having completed the same distance run, but over more and more difficult surfaces. The original line of runners has now been dispersed, and the delay between the first and last runner is called the dispersion measure.

How does this relate to pulsars? The line of runners at the beginning of the race represents a single pulse from a pulsar, and each runner represents a pulse at a specific radio frequency. The starting line is the pulsar and the observer at the finish line is here on Earth. When they search for pulses from a suspect pulsar, astronomers don't observe at a single radio wavelength, but over a section of the radio spectrum, or bandwidth. They do this in order to increase the sensitivity of the radio telescope so enabling them to find fainter pulsars. Despite the fact that the radio pulses at each wavelength leave the pulsar at the same time, by the time they reach the finish line they're all spread out – dispersed – and reach Earth at different times. From the astronomer's perspective, instead of seeing a single, sharp pulse, the pulse seems spread out over time, just like the stream of runners at the finish line of our obstacle race. And just like a stream of runners, the once individual pulse of radio waves is smeared into a constant radio hum.

But in reality it's even worse for pulsar astronomers; let's now see what astronomers really have to deal with. You see, in our rather unfair race we had a single line of runners all starting together, and we watched them all trail in after having completed their obstacle-ridden tracks. But imagine if the first row of runners was followed a few seconds later by a second line-up of equally fast sprinters identical to the first row, sprinters who also have to overcome the same debris-ridden tracks as the first line. Now let's look what happens at the finish line. No sooner has the last of the runners of the first line-up completed the race than the first of the second row completes his race, followed by the remaining runners in the second line-up. In fact the runners (pulses) can and often do overlap. What you would see is a stream of ten runners from the first race blending into a stream of ten runners from the second line-up. Now let's add a third line-up, then a fourth, and so on, each line-up leaving the starting blocks at equal intervals. As the last runner of each line-up completes the race – remember, they're all identical – the first of the following line-up reaches the finish line. What you see is not a sequence of runners completing a single race, but a constant stream of people. By observing this constant stream of people you would be unable to determine how much time had passed between one line-up starting the race and the next. To make matters even worse, the shorter the time between the start of each race, the greater the jumble of arrival times of runners, at least as seen from one end of the finish line. Thanks to dispersion and the rapid fire emission of individual pulses, the period and other details of a pulsar signal are difficult to determine.

How do astronomers get around this problem? On one hand, you could just narrow the bandwidth, the equivalent of looking only for runners with the number 1 printed on their shirts. But this would reduce the sensitivity of the radio telescope and many faint pulsars would be missed. The solution is to use a high bandwidth to maintain the sensitivity, but use a range of narrow-band receivers – devices that collect the signal from the antenna itself – to keep a record of the entire signal in discrete narrow wavebands. In a kind of reverse engineering, the signals coming out of each narrow band receiver can then be delayed different amounts in the opposite way to that expected from different dispersion measures. By delaying the observed arrival time of the signal at shorter wavelengths (runner 1) compared with longer wavelengths (runners 2, 3, and so on), the pulses in each consecutive pulse can be evened up to 'arrive' at the same time. This has the effect of reintegrating the once dispersed pulsar signal. Once the astronomers have done this, they can find the period of the pulsar and lots more besides.

As we saw earlier, the data recorded from a pulsar suspect has to be sampled at a higher rate than the pulsar period in order for the individual pulses and hence the pulse period to be identified. The sampling rate is limited by the dispersion measure: the greater the dispersion, the more each individual pulse will be smeared out, which means that higher sampling rates will be difficult. By compensating for dispersion the individual pulses can be seen. The problem is, how do you know at what rate to sample the data if you can't see the individual pulses? The only way to see the pulses is to de-disperse the signal, but how do you know how much dispersion to compensate for? In fact, astronomers use computers to analyze the signal from a pulsar suspect in both dimensions of dispersion measure and pulse frequency simultaneously. In the case of CP 1919 and the Vela pulsar, the astronomers were able to discover the periodic nature of the signal simply by speeding up the recording, thereby spreading out the signal and revealing the individual pulses. In the case of the Crab pulsar, the dispersion is so high that even a channel 100 kHz wide is smeared out over the 33 ms period. The discovery depended on the individual giant pulses, not on the periodicity.

Dispersion as an Astronomical Tool

So far we've viewed dispersion as an obstacle to learning about pulsars, but in fact astronomers can use dispersion to their advantage. Measuring the distance to all but the nearest stars is difficult, but it turns out that dispersion measure is a built in distance indicator. To return to our race analogy, the difference in arrival times of the runners depends not only on the degree of difficulty of their respective tracks, but on how long the track is: the longer the race, the greater the delay in arrival of the runners from 1 to 10. In the same way, the further away the pulsar and the more space it has traveled through, the greater the dispersion. Using a model of the electron density in the Galaxy allows an estimate of the distance to the pulsar. Knowing the distances to pulsars means you can make a three dimensional map of the distribution of pulsars in the Galaxy.

Parameter Space

All these factors influence how the search method is designed. Whenever searching for a signal, you need to define its characteristics and then devise a way of detecting it. Since you don't know the characteristics of any of the pulsars you are searching for, you need to search a wide range of possible values, or parameters, of each for every point on the sky you look at. This is known as parameter search space, and consists primarily of period, pulse width and dispersion measure. The only efficient way to do this is by using computers, and this is why they now play such a major role in pulsar surveys. Parameter space is immense, however, and so cannot be searched in real time. Unlike Hewish and Bell standing over the paper chart recorder waiting for the transit if CP 1919, pulsars are now discovered after the actual observations have been made by using computers to sift through recorded data, applying one set of parameters – and combinations – after another.

What Wavelength?

The choice of the wavelength at which to observe depends on a range of factors, especially the relation between the radio spectrum of the pulsars and the galactic background. Pulsars can be observed at anything from 15 m to 1 cm (20 MHz to 30 GHz), being considerably stronger at lower frequencies. However, just as fog can reduce visibility for drivers at night, so the gas and dust that pervades the Galaxy reduces the visibility of pulsars. And just as yellow light tends to penetrate fog better than white light, searching for pulsars at specific radio wavelengths improves the chances of detecting them. It turns out that at a wavelength of 20 cm (1.4 GHz) pulsars stand out much better against the background of the Galaxy. At this wavelength, interstellar scattering and dispersion are less, especially for pulsars near the galactic plane, and the galactic radio continuum is weaker.

Second Molonglo Survey

A number of searches for pulsars were conducted in the decade following their discovery, and the most important ones are listed in the box. But as a case study we will look at the most successful, the second Molonglo survey carried out in

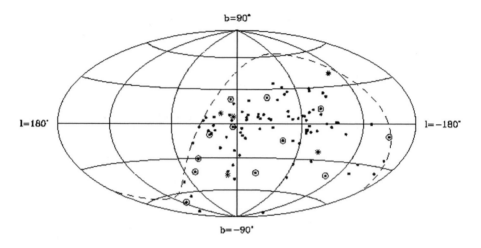

Figure 8b Results of the second Molonglo Pulsar Survey in which 150 new pulsars were discovered.

1977. The survey was led by Dick Manchester (who we met in Chapter 5). When pulsars were announced to the astronomical community, a young Manchester had just been given a job at the Parkes radio telescope. Manchester grew up in Canterbury on the South Island of New Zealand where he completed an honors degree in physics before starting a PhD. He grew up with a love of the stars thanks to his uncle, an electrical engineer with a passion for astronomy. But there were few opportunities to study astronomy at Canterbury and so he joined an ionospheric and meteor group lead by Dr C D Ellyett. 'A few months after I started, Dr Ellyett announced he had accepted the post of Professor of Physics at the University of Newcastle in Australia,' Manchester recalled. 'So the whole research team of about six people, with me as the junior member, up and left Canterbury for Newcastle.'

Manchester completed his Ph.D. in 1968, and it was then that his latent interest in astronomy began to surface. He contacted the CSIRO's Radiophysics Laboratory, which had been pioneering radio astronomy since the end of the war. An astronomer working at the CSIRO's Parkes Radio Telescope in western New South Wales, John Bolton, encouraged the young Manchester to pursue a career in astronomy. The most common next step for a new PhD is to gain a position at an overseas institution, but Bolton couldn't decide where overseas would be suitable for the keen but inexperienced scientist and expressed pessimism over Manchester's prospects. It seems Bolton may have had an ulterior motive because he eventually asked 'How would you like to come here?' It didn't take long for Manchester and his wife, Barbara, to realize that it was a golden opportunity. They moved to western New South Wales in a very hot February in 1968. 'As it turned out, the week I started at Parkes was the same week the *Nature* paper announcing the discovery of pulsars appeared in print. I

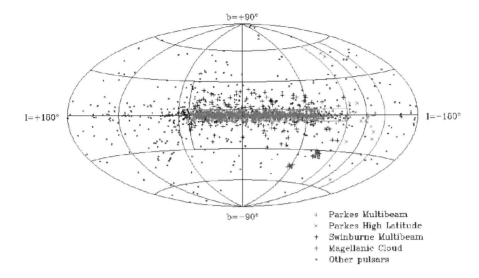

$b=+90°$

$l=+180°$ $l=-180°$

$b=-90°$

- Parkes Multibeam
× Parkes High Latitude
+ Swinburne Multibeam
+ Magellanic Cloud
• Other pulsars

Figure 8c Results of the Parkes Multibeam Pulsar Survey and other surveys.

wasn't especially conscious of it at the time. I was still finding my feet at Parkes ... I'd never done astronomy before.' As we saw in Chapter 5, it wasn't long before Manchester experienced pulsar astronomy first hand, with the discovery of the first pulsar glitch. But this was just the beginning.

In 1969 Manchester was offered a job at the National Radio Astronomy Observatory (NRAO) in Charlottesville in Virginia where he met his long-time friend and colleague Joe Taylor. Taylor was an Assistant Professor at the University of Massachusetts, and was already involved in pulsar research. Two years later, Manchester was offered a faculty job at the University, first as an Assistant Professor and later as Associate Professor. That same year Manchester met another of his long-term colleagues Andrew Lyne who was based at Jodrell Bank Observatory in England. Manchester returned to Australia in 1974, rejoining the CSIROs Division of Radiophysics. There was a growing need to find out just how many pulsars there were. Based on the tally of stars and the estimated frequency of supernovae in the Galaxy, there should be thousands of pulsars out there. While not all of them would be visible from Earth, 9 years after their initial discovery only 149 had been found. Where were the rest? Manchester decided he wanted to find out.

The trip from Sydney to Parkes in western New South Wales takes about five hours, and Lyne and Manchester had made the trip many times before. They used to plan a lot of surveys during those journeys, and it was during one such trip they came up with the idea of the Molonglo survey. The idea was to search for pulsar suspects using the relatively wide-field capabilities of the Mills Cross telescope (the same used by Large and Vaughan to discover the Vela pulsar), then follow each one up with the high resolution of the Parkes telescope. An important aspect of the survey was that it brought together several people who had particular expertise in areas that all came together very nicely. Michael Large knew the Molonglo system and came up with the multibeaming arrangement that was crucial to the success of the survey. Andrew Lyne brought expertise in electronics and built new front ends for the Molonglo telescope. These devices represent the first stage in the amplification of the signal from an antenna, the signal then being sent to wherever it is being processed. In the case of the Mills Cross that is half a mile from either end of the radio antenna to the center. One of the basic principles of engineering is that any loss between the receiver and the processor adds to the system noise and decreases the sensitivity. You can overcome this loss by putting in gain: the first stage amplifier is placed as close to the feed as possible. The original Molonglo telescope didn't have those amplifiers and so it had a fairly high 'system temperature' making it less sensitive than was needed for pulsar surveys. Lyne built the electronics and, with the help of the Molonglo people, installed them at Molonglo. The new front ends doubled the sensitivity and hence doubled the number of pulsars found. Lyne's contribution of new front ends were used long after the pulsar survey had finished, benefiting many unrelated research projects. Taylor was very good on signal processing. He understood fast Fourier transform processing and harmonic summing needed to analyze the signals in order to identify the pulsars. Finally, Manchester knew the Parkes telescope. He had skills in programming and set up the Parkes processing.

Parkes and Molonglo turned out to be highly complementary. Molonglo is a long, thin reflector – several times the collecting area of Parkes and hence has good sensitivity. But being more or less fixed to the ground it could only see objects as they passed overhead into its beam. The width of the beam in declination is about four degrees, and so as the sky rotates overhead the Molonglo telescope is looking at a four degree strip of each RA line that crosses its beam. Initially, this meant that the positions of any pulsar suspects were not well known so that follow-up observations were difficult. To get around this problem, Manchester wrote a program so that the Parkes telescope could scan down each RA strip which contained a pulsar suspect. As it did so, the signal was analyzed continually in real time around the search parameters until the pulsar showed up. Once the position of the pulsar had been identified, the astronomers could use the Parkes telescope to go back and observe it as often and as long as they liked. Using this technique, the astronomers surveyed the entire southern sky up to a declination of $+22°$.

The Molonglo stage of the survey lasted about a year, with the confirmation of the pulsar suspects at Parkes taking about six months. Parkes was then used to monitor the freshly discovered pulsars for a further year in order to refine each pulsar's characteristics. Because Parkes used multichannel receivers it was able to analyze the signals at a range of wavelengths immediately, making it more efficient at identifying pulsars with high dispersion measure. The data were recorded onto magnetic tape and then processed at the Radiophysics laboratory in Sydney.

By any measure, the survey was a raging success. Of a total of 224 pulsars identified in the survey – by itself an excellent sample for statistical purposes – 155 pulsars were new discoveries. The second Molonglo pulsar survey had doubled the number of known pulsars. In addition, the second known binary pulsar (PSR 0820+02) was discovered in this survey.

What the Surveys Revealed

The modern picture of the number and distribution of pulsars has been painted using data from many such surveys, each one contributing their own discoveries. Modern estimates place the number of pulsars in our Galaxy at 150,000, each one radiating for around 10,000,000 years. This means that, assuming the population remains steady, a new pulsar must form about every 70 years or so, a conclusion supported by the best estimate of a new supernova in our Galaxy every 100 years. Along the way, however, there were some spectacular discoveries, and as we'll see, pulsar astronomy was about to warrant the awarding of yet another Nobel Prize.

SUMMARY OF PULSAR SURVEYS

The following is a summary of the most important pulsar surveys conducted in the decade following the discovery of pulsars. Some of the more important features of each survey are included.

1st Molonglo Survey
Michael Large and Alan Vaughan (1971)
Ideal telescope to follow up pulsar discovery, with a large collecting area and hence high sensitivity.
- Had already discovered 9 pulsars, including Vela pulsar (described in Chapter 5).
- 31 new pulsars discovered (out of a world total of 58).
- Sensitivity down to 80 mJy.
- Visual inspection of chart paper similar to the way Bell found the original pulsars.
- DEC +20 to -90 coverage.

Jodrell Bank Survey
Joseph Davies, Andrew Lyne and John Seiradakis (1972, 1973)
- 39 pulsars discovered using the 250 foot Jodrell Bank telescope.
- Sensitive to periods between 0.08 and 4.0 seconds.
- 15 mJy.
- ~1 sr along galactic plane.
- Computer search for periodic signals.

Arecibo Surveys
Roger Hulse and Joe Taylor (1974, 1975)
- 40 pulsars discovered using the 300 meter Arecibo telescope.
- Periods sensitive from 0.03 to 3.9 seconds.
- Increased sensitivity (10 times more sensitive than previous surveys) down to 1.5 mJy.

2nd Molonglo Survey
Manchester et al. (1978)
- 155 pulsars discovered, making it the most successful survey to date.
- Molonglo and Parkes telescopes.

Greenbank Survey
Stokes et al. (1985) and Dewey et al. (1985)
- 54 pulsars discovered using the 300′ Greenbank transit parabaloid, West Virginia.
- Similar sensitivity and complementary to second Molonglo survey.

Jodrell Bank 1.4 GHz Survey
Trevor Clifton and Andrew Lyne (1986)
- 40 pulsars discovered using the 250 foot Lovell telescope at Jodrell Bank observatory.
- Most with large dispersion measures.

9 'Two by Two'

'If the neutron star hypothesis of the origin of Supernovae can be proved, it will be possible to subject the general theory of relativity to tests which according to the considerations presented in this paper deal with effects which in order of magnitude are large compared with the tests so far available. The general theory of relativity, although profound and exceedingly satisfactory in its epistemological aspects, has so far practically not lent itself to any very obvious and generally impressive applications. This unfortunate discrepancy between the formal beauty of the general theory of relativity and the meagerness of its practical applications make it particularly desirable to search for phenomena which cannot be understood without the help of the general theory of relativity.'

Fritz Zwicky (1939)

In the early 1970s, Dick Manchester was in the United States sharing an office with Joe Taylor at the University of Massachusetts. It was the beginning of a long professional relationship that survives to this day. While they collaborated on many projects, there was one that Taylor initiated that Manchester didn't take part in. Taylor proposed to use the largest radio telescope in the world, the Arecibo radio telescope on the island of Puerto Rico, which borders the Caribbean Sea and the Atlantic Ocean. In order to gain time on any professional astronomical instrument, astronomers need to submit a proposal outlining what they hope to achieve, hopefully justifying in the eyes of the telescope's managers[1] why valuable telescope time should be given to the astronomer submitting the proposal. The goal of Taylor's project was much the same as other pioneering pulsar surveys: push the limits of sensitivity and build up a statistical sample of pulsars in the Galaxy that would help astronomers understand them.

At the time, few took the possibility of finding a binary pulsar – a pulsar sharing an orbit with another star, either a normal star or a neutron star – seriously. It is true that the majority of stars in the Galaxy are in binary systems. However, considering their supernova origins, pulsars were assumed to be solitary creatures. Surely a supernova explosion would blow off so much mass from the exploding star that a binary system would fly apart, leaving the pulsar isolated in space? As attractive as the idea of a binary pulsar was, it certainly wasn't high on the list of astronomers' expectations. Nonetheless, when he applied to the National Science Foundation for funding to purchase a computer for the search, Taylor mentioned the possibility of finding a binary pulsar. It would be one of those unlikely but extremely important discoveries. Determining the mass of a single pulsar spinning alone in space is not easy; find one with an object of known mass and you have a chance to learn a great deal more about the pulsar. It would be highly significant, as Taylor pointed out, 'to find even one example of a pulsar in a binary system, for measurement of its parameters could yield the pulsar mass, an extremely important number'. No one knew that out of this search would emerge just such a discovery, one of the most important in the history of astronomy.

Investigations

An expert at deciphering pulsar signals, Taylor had already played a major role in other surveys. Taylor completed his PhD at Harvard University in January 1968 shortly before Hewish announced the discovery of pulsars. He immediately drew up a proposal with colleagues to use NRAO 92m telescope to study this new, bizarre phenomenon. It did not take long for him to make his mark in the fledgling pulsar community. Writing software capable of distinguishing pulsed, dispersed signals from background noise, Taylor discovered the fifth known pulsar (following the original four discovered by the Cambridge group) in June 1968. As we saw in the last chapter, there was a tremendous need to find more pulsars and develop a statistically valid sample set of pulsars. With this goal in mind, in 1972 Taylor initiated the pulsar survey using the 300 m Arecibo telescope. Arecibo was chosen since it was the largest telescope available for low radio frequency work. Although it is the largest telescope in the world, it has the limitation that it is not fully steerable. At 300 meters across, mounting such a dish so as to be movable is impractical, so it was built into a natural depression in the hills of Puerto Rico. A huge gantry is suspended above the dish from cables attached to three tall pylons. The radio receivers themselves are suspended at the focus of the dish and can be pointed to different parts of the dish, allowing astronomers to 'point' the telescope at different places in the sky surrounding the zenith. Combined with Earth's rotation, the massive Arecibo telescope is able to examine a 20 degree wide swathe of sky every 24 hours. Despite the limitation, the instrument's sheer size and sensitivity made it an attractive telescope for searching for pulsars.

But he needed an assistant. Taylor approached a young graduate student, Russell Hulse: would he be interested in carrying out a new, high sensitivity search for pulsars as his doctoral thesis research? The idea was attractive to Hulse since it combined three subjects Hulse was particularly interested in: physics, radio astronomy and computers. It would be Hulse's job to design the exact search method to be used, devise extremely efficient mathematical algorithms and programming techniques so as to allow a minicomputer to process the received signals, and then live on the island and operate the telescope along with the rest of the pulsar search system. Approval for the survey was duly given and Taylor and Hulse traveled to Arecibo. The goal was to maximize the sensitivity of

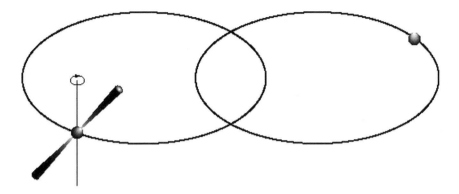

Figure 9 Binary pulsar (not to scale). The orbits of the pulsar and the normal star share a common center, and periodically pass close to each other. Illustration by the author.

the telescope by making the most of the signal processing, using the dedicated minicomputer that Hulse had programmed to analyze the signals. Observations were carried out at 430 MHz. The bandwidth of 8 MHz was split into 32 adjacent bandwidths each 250 kHz wide. Suspect signals were detected by the computer, sampled, de-dispersed and written to magnetic tape in real time. Hulse used a computer program called ZBTREE that did the de-dispersion in real time, with its output data on magnetic tape then later fed to a program called CHAINSAW, which looked in the de-dispersed data streams for the weak, periodic pulsed signals characteristic of pulsars. The analysis searched for pulsars with periods ranging from 4 seconds down to 33 ms. As we saw in the last chapter, finding pulsars involves searching parameter space. In fact there were five parameters to search through: pulse period, pulse width, dispersion, and two coordinates on the sky. The computer tried half a million parameter combinations for each 136 seconds of radio signals taken from a strip of sky 10 arc minutes across in each strip. From the beginning the survey was a success, with 40 new pulsars discovered at a time when only about 100 were known.

Fortuitous Down Time

At the time of the survey, the Arecibo telescope was undergoing a major upgrade. While this placed severe limits on many of the scheduled observing programs, it didn't interfere too much with the pulsar survey. Being a graduate student with few formal commitments, Hulse was able to live on site during the construction. Between interruptions caused by the upgrade, Hulse could use the telescope to search for pulsars, taking advantage of opportunities as they arose. This flexibility allowed Hulse to spend more time searching for pulsars than either he or Taylor could have anticipated. It also allowed him to experience something which is quite rare these days: great freedom in the use of a major scientific instrument, over an extended period of time. At the same time, there were many frustrations, such as putting up with radio interference, which is common at the relatively low radio frequency used for the pulsar search. Radio noise sources ranged from lightning in thunderstorms, faulty aircraft warning lights on the tops of the aerial supports to naval activities off the coast, which totally obliterated any feeble signals coming from outer space: 'I just sat in the control room watching signals from the naval radars (or whatever) jump around on the observatory spectrum analyser,' Hulse recalls. Operating this way, Hulse spent over a year at Arecibo, from December 1973 to January 1975.

PSR 1913+16

The computer was set up to report on any pulsar suspects above a certain sensitivity threshold. In July 1974, one of the pulsar suspects, PSR1913+16, came in at a strength just above the cut off; if it had been any fainter, it would have been missed by the computer altogether. What was exciting about this pulsar was its period: spinning once every 59 ms, it was the second fastest pulsar known after the Crab pulsar. Hulse's initial response was to write the word 'fantastic' at the bottom of the record. On August 25th 1974 Hulse made his first serious attempt to accurately determine the pulsar's period. The technique was to make two observations, each of between 5 and 15 minutes long, at either end of the 2 hour observing window afforded by the Arecibo telescope as it tracked a given object. These observations were then combined to provide an accurate period determined across the two hour observing interval. When Hulse did this for 1913+16, he came up with a result that was, in his words, 'completely perplexing': the two observations could not be combined in the usual fashion, as the periods measured within each of the two observations differed from each other by 27 microseconds. Although this may sound like a tiny difference, in the high-precision timing world of pulsar astronomy it was an enormous difference. He repeated the attempt two days later, but was presented with an even greater discrepancy. The period kept changing! Time after time, Hulse tried to pin down the period, which leapt about so much he kept crossing out and rewriting the period in his notebook, eventually rubbing them all out in frustration. This was no Eureka moment, but rather one of, 'What's wrong now?'

The sampling rate employed for the period determination observations was 16.7 ms. Now since PSR 1913+16 had a period in the region of 59 ms, the sampling rate was not really ideal. A 16.7 ms sampling rate provided poor resolution of the pulses from PSR 1913+16. In order to clarify the pulsar's behavior Hulse had to set up a series of special observing runs, including writing a special signal processing program for the Arecibo computer just for this set of observations, to more satisfactorily analyze the signal. Hulse carried out further observations on September 1st and 2nd 1974, expecting the problem to go away, but it just became worse. The period problem remained, but at least Hulse had eliminated one important possible source of error in the observing technique.

Hulse's observations enabled him to produce two graphs showing how the pulsar's period varied over time. Each graph showed a downward curve, as if during each observation the pulsar was speeding up. Intriguingly, the curves were identical, except that they were at different times. The fact that the curves were identical strongly suggested an instrumental error, but Hulse could not imagine what or where that error might be. The breakthrough came when Hulse realized the curves would match exactly if he shifted them about 45 minutes. It didn't take him long to realize that the period variation was not only real, but that it could only be caused by one thing: Doppler shift of the signal from a pulsar in a binary system. Somewhere along the line, the period had to increase again, to repeat the cycle as the pulsar completed each orbit. 'I immediately focused my observing schedule on getting as much data on PSR 1913+16 as I possibly could, as fast as I possibly could,' Hulse wrote. He was conscious of the importance of the discovery, but wanted to be absolutely sure he was correct before making any announcements, so he decided to wait until he had observations showing the increasing period as evidence of the binary pulsar.

Two weeks would pass before Hulse had an opportunity to detect the final piece of evidence supporting the binary pulsar discovery. On September 16th 1974, he was able to produce the desired result: a curve that showed the pulsar period falling 70 microseconds in less than two hours, reach a minimum, then begin to increase again. It was now confirmed: PSR 1913+16 was in a high velocity binary orbit with a companion star, with a period of about 8 hours. Two days later he had analyzed the data and wrote to Taylor who was teaching back at the University of Massachusetts, telling him of the discovery. This seemed inadequate, and phone connections out of Arecibo were difficult. With a sense of urgency he contacted Cornell University by shortwave radio, which linked to its counterpart in Ithaca, New York, where a secretary could connect the radio to telephone for the call through to Taylor at the University of Massachusetts in Amherst. Unsurprisingly, Taylor was quickly on a plane to Arecibo, taking with him equipment which could more efficiently analyze the pulsar's signal than the system Hulse had patched together. Taylor stayed at Arecibo for a short time helping Hulse gather more information on the pulsar. After Taylor went back to Amherst he remained in contact with Hulse via shortwave radio, with Hulse reading out long lists of numbers on the pulsar, which Taylor then used to analyze the binary's orbit.

So why did Hulse's original observations seem so perplexing? An initial analysis of the pulsar showed that it was in a close elliptical orbit with an unseen companion. In the months to follow it would become clear that the companion was also a neutron star, although no radio pulses were visible from it. By pure chance, the orbital period is close to eight hours, or one-third of a day. Using the Arecibo telescope, Hulse was constrained to observe the pulsar only once every 24 hours, by which time the pulsar reached almost the same place in its orbit (having completed three orbits of just less than eight hours each). The precise orbital period is 7 hours 45 minutes, 15 minutes short of an even multiple of 8 hours. After having completed 3 orbits, the pulsar was 45 minutes out of synch with the previous day's observations, explaining why Hulse's curves were separated by about 45 minutes. The 45 minutes was simply the accumulated difference of three complete orbits of the pulsar. If Arecibo was a fully steerable instrument, Hulse would have been able to track the pulsar for longer, and there would have been much less mystery to resolve. Nonetheless, had it not been for the sensitivity of the Arecibo telescope, and the opportunities afforded by the upgrade work, Hulse might never have made the discovery.

The Potential of PSR 1913+16

Immediately upon measurement of its orbital parameters, it was clear that the binary pulsar was more than just a tool for determining the mass of a pulsar, but rather had the potential to be a laboratory for studying the effects of general relativity. Let's think about this system for a minute. It has two high mass objects orbiting one another in a very eccentric (that is, elongated) orbit. Subsequent observations have revealed that the pulsar has mass 1.441 solar masses, while the other weighs in at 1.387 solar masses, strongly suggesting that both are neutron stars.[2] At any rate, Taylor and Hulse's observations showed that at closest approach the two stars were passing within a distance less than the diameter of the Sun, making it clear that the companion was also a compact star, most likely another neutron star. It is noteworthy that these masses are so close to the limit set by Chandrasekhar so many years before. During each orbit the two neutron stars fall in towards each other at up to 300 kilometers a second, ten times the velocity of Earth around the Sun. Such speeds are regarded as relativistic, because they represent a sizable fraction of the speed of light, in this case about a thousandth. At some point the Doppler shift becomes zero as the pulsar is neither approaching nor receding from us. The two stars then swing quickly around each other before arcing back out into space, repeating this sling-shot every eight hours. The celestial dance is visible from Earth as a rhythmic change in the pulsar's period. The pulse arrival times vary by a few seconds due to the Doppler effect: when the pulsar is approaching its pulses arrive more frequently; as it recedes they are more widely spaced. An equivalent way of looking at this is that the apparent pulse period is altered by the fact that the radio signal has further to travel when the pulsar is furthest from us, and has less distance to travel when it is nearest. The rapid orbital velocity means that the system completes about 1,000 orbits per year, or 20,000 orbits in 20 years. Since the pulsar represents an ultra-accurate clock in a system operating at relativistic speeds, and in a strong, time-varying gravitational field, the binary pulsar is a dream come true for physicists interesting in studying general relativity. In order to appreciate the importance of this discovery, we now need to take a detour and explore the nature of spacetime.

Understanding Curved Space

One of the most difficult things to understand about space is that it is, in a sense, a substance that can be bent. This was described most completely by Albert Einstein in his General Theory of Relativity. One of the early triumphs of general relativity was the explanation of the advance of Mercury's 'perihelion'. As Mercury or any other planet moves around the Sun, its elliptical orbit carries it closer to the Sun and then further away. The closest approach is called perihelion. It was a perplexing problem that the position of Mercury's perihelion continually shifted in space. The shift was tiny – a mere 43 seconds of arc every 100 years – but it was enough to drive the theoreticians to predict the existence of yet another planet orbiting the Sun inside of Mercury's orbit, the mythical planet, Vulcan. Despite searches, no extra planet could be found. General relativity solved the problem by explaining that space was not a flat, inflexible matrix through which the stars and planets moved, but rather a flexible, although extremely stiff, 'fabric' which could be bent and distorted by anything with mass. While Newton had theorized that gravity was an attractive force between any two objects with mass, Einstein showed that gravity was a distortion in the fabric of spacetime. In the case of the Solar System, the Sun was creating a depression in spacetime, much like a bowling ball in the middle of a large rubber sheet. The planets, while creating their own much smaller depressions, followed the curvature in spacetime created by the Sun.

One way to measure the curvature of a surface such as a sphere is to measure the perimeter of a circle and compare it with the surface area marked off by the circle. The same goes for an ellipse or any other shape. By comparing how the area of an ellipse in curved space differs from one in flat space, we can measure how curved the space is. You can illustrate this point by drawing an ellipse on a soccer ball and an ellipse of the same size on a flat piece of paper. The ellipse drawn on the soccer ball has a greater area because of the curvature of the ball's surface. The same difference in area applies to the space marked out by the orbits of the planets. Because the space that Mercury moves through so near the Sun is so steeply curved, the ratio of the perimeter of its orbit and the area traced out by the orbit is noticeably different from the same ratio in Newtonian flat space. When you measure the area traced out by the planet based on the distance it has traveled around the perimeter of the orbit with the area based on the curved space, they don't match. The area in curved space will always be a little more.

In reality, what happens is that at the end of every orbit, Mercury has a little bit of its journey through space left over due to the curvature of space. To make up for this, at the 'end' of each orbit Mercury has to overshoot the mark, traveling further in its orbit than it would if it were in flat space. As a result, the entire ellipse tends to twist a little on each orbit of the planet – the perihelion point where the planet is closest to the Sun moving around just a little each time – so that the orbit traces out a pattern that looks like the petals of a daisy. This movement of the point of perihelion is called precession, and all of the planets do it. When the curvature is slight, as for the planets far from the Sun, the difference is minuscule. By the time you take into account all of the various gravitational tugs and pulls exerted by the other planets the perihelion precession of the outer planets is completely swamped by these larger effects. For tiny Mercury deep within the gravitational well created by the Sun, the difference is significant. Mercury's perihelion precesses by 43 seconds of arc – a fortieth of the diameter of the full Moon – every 100 years. At this rate, it will take Mercury's perihelion point four million years to complete one rotation of the Sun. The Earth, incidentally, does the same thing every 30 million years. Small as these figures might be, they are enough to confirm Einstein's theory of general relativity.

Now let's consider the binary pulsar PSR1913+16. I described earlier how the orbit could be mapped in detail by watching how the period varied due to the Doppler shift. If space was flat then the orientation of the orbit would remain stationary. This would mean that the perihelion of the pulsar's orbit – that is, the shape of the velocity curve observed by Hulse at the time of discovery – would not change with time. However, space surrounding these massive objects is not flat, and the orbit does rotate for the same reason that Mercury's perihelion does. In fact, general relativity predicts that the orientation of the binary pulsar orbit would rotate a whopping 4 degrees per year. In a hundred years the orbit of the binary pulsar would have completed more than one complete rotation! In addition, as the pulsar's beam passes close to the companion, it passes through space that has been curved by the companion, further delaying the arrival time of the pulse beam by a small amount, and at closest approach, the pulses have to climb out of a deeper gravitational well created by both stars rather than just the pulsar, further redshifting and delaying the beam. These effects have all been studied in great detail in the two decades since the binary pulsar's discovery.

Ripples in Spacetime

But there was more to come. The eccentric orbit of the system means that it must be emitting a strange phenomenon called 'gravitational waves'. Just as spacetime can be curved, violent events such as supernova explosions or the merging of black holes can send ripples in spacetime radiating into the Universe. I said before that spacetime was stiff, and what I mean is that it takes a lot of energy to make it vibrate, that is to send out gravitational waves. The most effective source of gravitational waves will be one where the spacing between the masses changes from nearly zero to some maximum value. [3] Two masses joined by a spring would oscillate in this way, but two masses joined by a rigid rod and spun like a band leader's baton is much better. Viewed side-on, such a rotating dumbbell appears to expand and contract from maximum to zero. There's a problem, however. There's a limit to how fast something can rotate before it flies apart under centrifugal force. Even a steel ball bearing will explode if it rotates faster than a million times a second. We want to take two large, joined masses and rotate them as fast as possible. We've seen, though, that the bar that joins them will break at high speeds, and so it's better to use a solid elongated mass. Let's be ambitious and use 1,000 tonnes of steel, or even better a 10,000 tonne nuclear submarine. If we mounted the submarine on a turntable and rotated it to just short of its breaking point – say 10 times a second – how much gravitational wave energy would we produce? Sadly, not very much: about 10^{-24} watts. To put this into perspective, the smallest ant walking fast up a wall uses 10^{-7} watts, a billion billion times more energy than the gravitational waves produced by our rotating submarine. And the ant's efforts are 10 billion times smaller than an average family car. Although astronomers are used to detecting low-energy phenomena, such a tiny ripple in spacetime is far beyond the sensitivity of current detectors.

But there are much larger oscillating masses out there in space. Binary stars, for example, behave like sun-sized dumbbells. In our Galaxy there are millions of such systems. Each half of the binary system creates a full in and out motion of the masses, as seen edge on, so the gravitational wave frequency will be double the orbital period. This means that the gravitational wave frequency will be in the range of one cycle every few hours to one cycle every few days. Each individual source will be very weak and even the combined effect of all of them would be completely undetectable by Earth-based detectors, where disturbances such as temperature variations and tides cause relatively enormous variations on the same time scale as the signal that would totally swamp any gravitational waves from these sources.

Gravitational Waves Confirmed

As it turned out, PSR 1913+16 offered an alternative way to confirm the existence of gravitational waves. The gravitational field in the system is ten thousand times stronger than that felt by Mercury as it orbits the Sun. According to general relativity, at closest approach, the system should be emitting gravitational waves more than at any other time in its orbit. Now this means a loss of energy in the system, which will in turn cause the two stars to spiral closer together, in the same way that an Earth satellite will spiral in towards Earth if it slows down too much due to atmospheric drag. As they move closer together, the orbital period is predicted to decrease by 75 microseconds per year. In December 1978 Joe Taylor announced at a Munich conference that he and his partners Lee Fowler and Peter McCulloch had determined that the binary pulsar PSR 1913+16 is doing precisely this. By 1983 the measurements of the orbit yielded decay in the period of 76 microseconds per year. They had confirmed the emission of gravitational radiation exactly as predicted by general relativity.

As the binary pulsar continues to spiral in, it will emit ever stronger gravitational waves, losing energy at an ever faster rate and slowing down even more. In about 300 million years the two neutron stars will coalesce, merging to become one. Three years before they finally merge they will be screaming around each other at a thousand kilometers a second, completing an orbit every 3 seconds, and passing within a thousand kilometers of each other. A minute before the end the stars will be orbiting each other 15 times a second. Near the end they will be a mere 30 kilometers apart and orbiting 500 times a second at a sixth the speed of light. Only in the last few milliseconds will the final merger take place. As they come extremely close to each other, even something as stiff as a neutron star will start to become distorted, and eventually the two will likely merge to form a black hole. But as they merge they will briefly emit the gravitational wave luminosity of a hundred thousand galaxies. Observing the merger of this particular binary pulsar is out of the question for present day gravitational wave experimenters because it will not occur for many millions of years; however, they are very interested in the possibility that another such system could be nearing its end right now, and hence be observable.

Hulse and Taylor's discovery of the binary system PSR 1913+16 was a major event in the history of pulsar astronomy. It was more than just one more rapid pulsar, or even the first binary pulsar. It provided astronomers with a laboratory for testing different theories of gravity, of which Einstein's continues to reign supreme. It should also be noted that this particular binary pulsar, being such a speedy example, made it very difficult to detect. In the years to come other binary pulsars would be discovered, most with much more circular orbits. Manchester discovered a much longer period binary pulsar taking nearly four years to complete one orbit. This created the reverse difficulty encountered by Hulse, in that the binary nature was not immediately apparent and required a much longer series of observations to confirm. Yet another binary pulsar, PSR 0655+64, has an almost circular orbit with a period of almost exactly 24 hours. This presented similar problems for its discoverer, Damashek, as Hulse encountered, right down to the use of another transit telescope, this time the NRAO Greenbank telescope.

As astronomers extended the sensitivity of pulsar surveys, and especially now they knew what to look for, there was no doubt that many other binary pulsars would be found. By mid-1988 a total of nine binary pulsars had been discovered; by 2006 well over a hundred had been found. Some of these involve orbits with normal stars, others with brown dwarfs. At least one system would be found in which the pulsar plunges through a disk of material surrounding the companion star. In 1988 astronomers found a binary system dubbed the 'Black Widow' pulsar because it is slowly destroying its companion. Then, in 2006, the youngest binary pulsar system was discovered using the Arecibo telescope and confirmed using data from previous survey at Parkes. The importance of the 112,000 year old system is that its existence implies binary pulsars are being formed at a rapid rate in the Galaxy. Since pulsar lifetimes are in the region of tens of millions or years, the chances of astronomers stumbling across one so young are remote unless they are being formed all the time.

There is no doubt that it is the combined observational and theoretical skills of Taylor and Hulse that allowed such groundbreaking discoveries to be made. But there was another enigma to be solved. As we've seen with the Crab pulsar, pulsars slow down as they age: young pulsars spin more rapidly than old ones.

What's also important when it comes to determining their age is that the rate at which they slow itself declines over time: faster pulsars slow down at a faster rate than older pulsars. The stability of PSR 1913+16 is remarkable, however, since it is slowing at a rate 50,000 times slower than the Crab. This implies that PSR 1913+16 is immensely old, and yet it is spinning rapidly like a new born. This suggested that the old pulsar was somehow given a new lease of life by being spun up by accreting material and angular momentum from its unseen companion. As we shall see in the next chapter, this same phenomenon is in fact common in the Galaxy and has given birth to an entirely new breed of pulsars, objects that owe their phenomenal speeds to their companions which were once just ordinary stars.

References

1. Specifically, a 'time allocation committee' that determines who gets to use the telescope, for how long. Sometimes optical astronomers find this frustrating since their allocated observing time is determined often months in advance. If it happens to be clouded on the evening of their observations, too bad.
2. A 'normal' star of this mass would have been torn to shreds by the forces experienced in this system.
3. The following explanation of gravitational waves was published in *Ripples on a Cosmic Sea* by David Blair and Geoff McNamara (Allen & Unwin, 1997).

10 'Faster'

By the early 1980s, pulsar astronomy was a hot topic, but it was about to get hotter. In the early pulsar searches, the detectors used to search for pulsars sampled the incoming signal about every 20 milliseconds. This meant that pulsars with periods less than about 100 milliseconds were difficult to detect. This wasn't considered a problem because all the theoretical and observational evidence suggested that pulsars spin several times per second. For example, the Crab pulsar was the fastest pulsar then known with a period of 33.1 milliseconds; it was also the youngest, being the result of the famous supernova explosion observed by the Chinese a mere thousand years ago. Even the binary pulsar discovered by Hulse and Taylor – the second-fastest pulsar on record – had been given a new lease on life through interaction with its unexpected companion and so somewhat of an oddity. No, the future lay in the discovery of more pulsars, but faster ones were not really thought likely. But as is usual in this game, deep in the darkness was lurking a pulsar spinning so fast it made all others look almost pedestrian. It was simply waiting for someone to look.

An Odd Little Radio Source in Vulpecula

In 1978 an American astronomer from the University of California, Berkeley, Don Backer, was exploring how prevalent interstellar scattering of radio waves was near the plane of the Galaxy when he discovered an odd little radio source called 4C 21.53.[1] What caught Backer's attention was the fact that the source was compact enough to produce interplanetary scintillation despite the fact that the gas and dust close to the plane of the Galaxy should suppress such a point-like radio source. Further, no one had found a pulsar nearby despite sensitive searches, including that by Roger Hulse and Joe Taylor four years earlier during which they found the now famous binary pulsar (in fact, not far from 4C 21.53 which Backer was studying). The following year Backer was searching the literature for references to this radio source when he noticed a different radio source had previously been detected remarkably close by. Known only by its coordinates, the source called 1937 +215 was some 30 seconds west of 4C 21.53, close enough that the two could be associated if it was assumed the position of 1937 +215 was only slightly in error. But there were problems linking the two. For one thing, their spectra were entirely different, indicating two different sources of radiation. Secondly, at 60 arc seconds across, 1937 +215 was huge compared with the point-like source needed to produce the scintillation seen in

4C 21.53. Nonetheless, Backer hypothesized that these two represented a large, faint supernova remnant and its pulsar.[2]

A number of searches for a pulsar at the position of 4C 21.53 was subsequently conducted by different astronomers using a range of telescopes, but all failed to find any telltale radio pulses. Then in September 1978, William Erickson was using the Very Large Array in New Mexico when he discovered a radio source this time to the east of 4C 21.53. All the indications suggested that this new source was in fact the scintillating source Backer had seen; there seemed no need now for a pulsar-supernova remnant association. The case seemed closed. Then in 1980, Erickson reported that in fact both the west and east radio sources were showing signs of interplanetary scintillation. There was sufficient uncertainty in the position of the source 1937 +215 that two years later Backer and Miller Goss used the Westerbork Synthesis Radio Telescope in the Netherlands to confirm that the broad radio source 1937+215 and the scintillating source 4C 21.53 were in fact in the same position on the sky. This led to a search for pulses with periods down to four milliseconds, almost ten times shorter than the fastest pulsar then known, the Crab pulsar. Still nothing was found. Just a steady, point-like radio beam.

Faster

Four years later at the 1982 meeting of the International Astronomical Union in Patras, Greece, Backer still suspected a pulsar was lurking deep inside 4C 21.53. During that meeting, the riddle of this object was given high priority, but no conclusions were drawn. When he got back to Berkeley, Backer planned yet another search for pulses from 4C 21.53, this time with Berkeley student Shri Kulkarni. They intended to use the Arecibo telescope to search the object for pulses over a wide range of pulse periods and dispersion measures. Observations were made later that year and then, in September, Kulkarni and Mike Davis recorded the first tantalizing signs of a pulsar while sampling at a rate of once every 0.5 milliseconds. They checked the observations by pointing the telescope away from 3C 21.53 and, sure enough, the pulses disappeared. But the pulses were only seen for a part of the observation, and a few days later couldn't be found at all! The mystery had not yet been resolved. In November, Backer, Kulkarni, Davis and Carl Heiles took another look at their pulsar suspect, this time looking at periods down to an unprecedented resolution of a tenth of a millisecond. Straight away they saw the source was not only flashing on and off but also scintillating. It was then Backer realized that the reason that the pulsating source was intermittent the previous September was because of interstellar scattering by the material in the galactic plane. The scintillation that had attracted his attention in the first place was masking the pulses. They had found their pulsar at last. But this was no ordinary pulsar: this one was rotating once every 1.558 milliseconds! Spinning at a phenomenal 642 revolutions per second – 20 times faster than the Crab pulsar – the first millisecond pulsar had been found. It was to remain the fastest pulsar known for the next quarter of a century.

More Problems

The discovery of millisecond pulsars created even more problems for astronomers still trying to understand how 'normal' pulsars work. Here were objects spinning at up to 90% of the speed needed to break them apart. Just how the neutron stars managed to get up to that speed was an even bigger problem. When most normal pulsars are first formed, they rotate about a few tens of times a second. At this stage they have strong magnetic fields, a direct result of the core collapse during the supernova phase: as the core collapses, the intensity of the star's magnetic field increases proportionally. Both the strength of the magnetic field and the spin rate of the pulsar decline over the following few million years. Because young pulsars have strong magnetic fields, they slow down faster than older pulsars with weaker magnetic fields. This helps astronomers estimate pulsar ages – the faster they're slowing down, the stronger their magnetic fields, and the younger they must be. Conversely, if a pulsar is slowing at a leisurely rate, then it must have a weak magnetic field and hence be old. The discovery of millisecond pulsars bucked this trend: spinning up to a thousand times a second indicated youth, yet their slow decline indicated immense age. Clearly, millisecond pulsars were different from those previously encountered.

Making Millisecond Pulsars

There are a number of theories explaining how to make a millisecond pulsar, but the most successful involves a binary star system. It was already known that Hulse and Taylor's binary pulsar had been spun up by the transfer of material and angular momentum from the companion star. Could such a phenomenon explain yet higher rotation rates? Perhaps a closer look at how this might work is needed here. Imagine a binary system consisting of a massive star and a low mass star. The massive star evolves faster, passing through the red giant stage and, finally, a supernova. The remnant of this once giant star is a tiny yet massive neutron star. If the binary system remains intact, the system will now consist of a normal star sharing an orbit with a neutron star, perhaps even a pulsar. Over time, the pulsar slows down as it loses rotational energy; as the period increases, the signal strength becomes weaker. Finally it becomes invisible. Now the system consists of a 'normal' star sharing an orbit with an invisible companion. It's important to realize that while the neutron star is small and essentially invisible from Earth, its mass is still high, being around the mass of a sun. This is the reason why the two objects share an orbit, rather than one orbiting the other.[3]

Now the gravitational fields around these two stars overlap. Imagine you were traveling in a spaceship from one star to the other. When close to the first star,

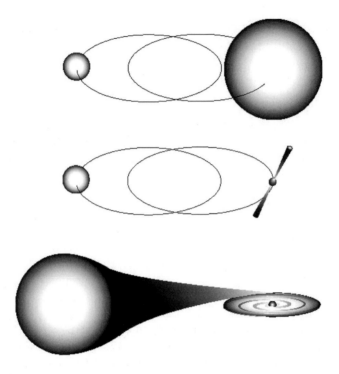

Figure 10 How to make a millisecond pulsar. In a binary star system, the larger star reaches old age first, explodes as a supernova and leaves behind a 'normal' pulsar, which eventually fades. In that time, the smaller companion evolves into a red giant, and matter spills onto the dead pulsar. This spins the old pulsar up, giving it a new life of incredible speed. Radiation from the new millisecond pulsar erodes the companion star which leaves a trail of matter that can eclipse the pulsar. Illustration by the author.

let's say the main sequence star, its gravity would dominate and you would have to spend all available rocket power to pull away from it. As you moved further from the star, however, the gravitational effects would weaken in accordance with the inverse square law: double the distance from the star and the gravity you feel weakens by a factor of 4, triple the distance and the gravity weakens by 9, and so on. But since there are two stars, and you were piloting your ship towards the second star, there would come a time when you felt not only the decreasing gravitational influence of the first star, but also the attraction of the second star. In fact you would reach a point where the gravitational effects of the two stars precisely cancel out, kind of like being on the peak of a ridge separating two roughly circular depressions. Move to one side of the ridge and you would tend to fall into one depression; move the opposite direction and you'd slide into the other. I say roughly circular because the presence of a second gravitating source distorts the space around the other, so that the curvature of space around each is shaped somewhat like a tear-drop, with the pointed ends of each just touching. A line joining equal gravitational influence from both objects defines the Roche lobes and is shaped like a figure 8.

Over time, the original pulsar 'dies': that is, its period is so long and the magnetic field is now so weak that the characteristic radio pulses no longer exist. By this time, however, the main sequence star will have evolved into a red giant. The expanded red giant star fills the curved space around itself, reaching out in all directions including towards the solar mass neutron star. Now even though the neutron star is no longer 'pulsing' it still has all the mass it began with, which is a lot, and it continues to distort the space around itself. Matter from the red giant fills its Roche lobe and spills into the gravitational depression created by the neutron star. To return to the analogy of two adjacent depressions, it is almost like one filling with water: eventually the water will fill the depression and spill over into the second depression. Matter begins to flood into the space around the neutron star. But the matter doesn't fall directly on to the neutron star; it spirals in gathering speed and momentum along the way. The matter forms a disk of material around the neutron star. Powerful magnetic fields surrounding the disk and the star transfer angular momentum from the disk to the neutron star. The ancient orb is given an extra spin the way some people can keep a basketball spinning on one finger by giving the ball an extra glancing push with their other hand. The in-falling matter causes the neutron star to pick up speed once more, only this time it spins well past its original rotation rate. The once dead pulsar is recycled to become a millisecond pulsar. The amount of matter needed to do this isn't much, only about 0.1 solar mass and the whole process takes between 10 and 100 million years. The rejuvenated pulsar can look forward to a much longer existence in its second incarnation. Because the magnetic field surrounding the millisecond pulsar is weak, it has little to cause it to slow down again and so can last for billions of years.

The scenario is supported by a number of observations. For example, as mass is transferred onto the pulsar, the matter is accelerated to tremendous speeds. As the matter hits the surface of the pulsar, it can reach 10 million K, emitting X-rays as a result. This is just what is observed in objects like Scorpius X-1, and such systems are known as Low Mass X-ray Binary Systems. Even the few millisecond pulsars not in binary systems can be explained. Single millisecond pulsars are thought to be the result of binaries involving two high mass stars. During the supernova explosion of the second star, mass loss can weaken the gravitational attraction between the two stars. If more than half of the total mass in the system is lost the two stars will go their separate ways. Alternatively, the intense radiation from the newly made millisecond pulsar can destroy what's left of the low mass star after accreting most of its mass.

Searching for Millisecond Pulsars

The galactic population of millisecond pulsars was once thought to be relatively low compared with normal pulsars. In 1988, however, Kulkarni and Ramesh

Narayan analyzed the results of a number of searches for millisecond pulsars to find out how many existed. Their conclusion was that the Milky Way plays host to up to 100,000 millisecond pulsars. But this was based on an analysis of the three then known millisecond pulsars. Where were the rest? One of the earliest surveys was carried out at Parkes. Observations began in May 1991 when only 8 disk millisecond pulsars were known. The survey was a cooperative effort between astronomers from the Australia Telescope National Facility, Jodrell Bank, and the Instituto di Radioastronomia CNR, Italy. The Parkes telescope was used to observe a total of 45,000 points on the sky, each observation lasting about two and a half minutes. In order to detect millisecond pulsars, the data were sampled 3,000 times a second. The quantity of data produced was too great to analyze at the telescope, and so the observations were stored on half inch video tape and later searched for millisecond pulsars. The survey was a success. Not only did the researchers double the number of known galactic millisecond pulsars, they also made some extraordinary discoveries.

One of the most important is the closest millisecond pulsar to the Earth, PSR J0437-4715. According to Dick Manchester who headed up the survey, the closest millisecond pulsar is really quite amazing. 'It's enormously strong,' he told me. 'It really knocked us over when we discovered it because most of these things are pretty weak and you've got to look at them for quite a while before you see too much. But this one just booms in.' PSR J0437-4715 lies about 150 parsecs away. One advantage of having such a bright pulsar is being able to detect individual pulses. Its period is now known to 15 decimal places. The pulse rate can be interpreted as a musical note, explained Manchester: 'You can, in fact, hear it if you connect head phones to the receiver output. It's E-flat so I'm told, although I'm not a musician.' The excitement of discovering the nearest millisecond pulsar didn't have time to die down before the faint image of its companion was found on photographs taken with the ANU[4] 2.3 meter telescope at Siding Spring Observatory in NSW. Later observations with the 3.9 meter Anglo-Australian Telescope revealed the companion to be a white dwarf. The pair are in an almost circular six day orbit.

Millisecond pulsar surveys continue around the world, including those carried out at Arecibo (using an ATNF-built multibeam instrument) of the galactic plane, as well as surveys carried out using the largest fully steerable radio telescope in the world, the 100 meter Robert C Byrd Greenbank Radio Telescope in West Virginia, USA (the successor to the telescope used by Staelin and Reifenstein to discover the Crab pulsar in November 1968). Surveys are also underway using the Giant Meter Wave Radio Telescope, an array of thirty, fully steerable 45 meter radio dishes built about 80 kilometers north of Pune in India. The large collecting area of this instrument could lead to a quadrupling of the number of pulsars known in the Galaxy.

Gravitational Beacons

Millisecond pulsars are not only interesting in themselves as an example of the unexpected directions stellar evolution can go, but also lead on to other research areas. We have already seen how Hulse and Taylor were able to demonstrate the emission of gravitational waves by studying the binary pulsar system PSR 1913+16. Now pulsars may hold the key to making the first direct detection of gravitational waves as they move through the Universe. The problem with gravitational waves, as we saw earlier, is that they are incredibly weak and so difficult to detect. But the discovery of so many millisecond pulsars in all directions in the sky gives astronomers an opportunity to sense these ripples in spacetime as they pass through the Solar System. Millisecond pulsars are almost perfect clocks, and so any variation in their period is not due to changes in the pulsar itself, but rather in changes associated with the observer. By observing a large number of them using one or more telescopes, astronomers using the Parkes telescope hope to set up the Parkes Pulsar Timing Array. Gravitational waves have the effect of changing the shape of a mass in two directions at once: as one dimension expands, the perpendicular dimension contracts, with the relative changes reversing as the gravitational waves pass through. Having a number of millisecond pulsars under continual surveillance means that if a gravitational wave were to pass through the Solar System, the changes it would make to Earth would be detectable as a change in the observed period of a millisecond pulsar. And so such a timing array might allow the observation of phenomena that create gravitational waves, including the origin of the Universe and the merger of supermassive black holes in the centers of galaxies.

Remembering that millisecond pulsars owe their very existence to close encounters with another star, it is not surprising that in order to find more of them a logical place to look would be in regions where there is a high density population of stars. There is one type of astronomical object where stars are more than just crowded together, they are positively rubbing shoulders. It should come as no surprise, therefore, to learn that the most millisecond pulsars were discovered in these dense congregations of stars, and it is here that we will journey in the next chapter.

References

1. Discovered during the fourth Cambridge survey in 1965.
2. Ironically, he wasn't the first to make the connection: in 1965 Tony Hewish and his student Samuel Okoye submitted a paper suggesting that 1937+215 was a pulsar-supernova pair not unlike the Crab Nebula, but the paper was rejected with the referee's comment: 'too speculative'.
3. Strictly speaking, any two orbiting objects share a common center of gravity, including the Moon and Earth: it is just a matter of degree whether one is regarded as the center of the system or not.
4. Australian National University.

11 'Globular Pulsars'

Introduction

Look at most introductory books on astronomy and you're bound to find a picture of the Milky Way galaxy. This is our local 'island in space', a vast agglomeration of some 200 billion stars isolated from the rest of the universe in all ways except through that most feeble yet far reaching of forces, gravity. Often you will see a photograph of the nearest large galaxy to our own, the great Andromeda galaxy, presented as an analog of ours. But the Andromeda galaxy differs from ours in two ways. Firstly, it is roughly twice the size of the Milky Way. Secondly, it is a typical spiral galaxy, whereas the Milky Way is a barred spiral, these names taken from the most obvious patterns traced out by the brightest stars. In any case, the overall shape of the Milky Way galaxy is typically presented as a disk with a central bulge, like two fried eggs back to back. The yolk of eggs represents the central bulge of the Galaxy, while the surrounding disk is laced with beautiful luminous swirls that trace out the spiral arms. It is within these spiral arms that the vast majority of current star formation takes place, fed by an enormous supply of hydrogen gas. Star formation itself is triggered by 'density waves', the galactic equivalent of a traffic jam, that drift through the disk of the Galaxy bunching up the hydrogen gas clouds and causing them to collapse and fragment, eventually spawning clusters of fledgling stars. The true shape of the Milky Way, however, is not a disk but a sphere. Surrounding the visible disk is a spherical halo of dark matter that controls its bulk behavior. This space is also populated by much smaller orbs, glowing spherical collections stars. These are the globular clusters. There are perhaps three hundred of these wonderful objects orbiting the Milky Way, each one a collection of anywhere from a few thousand to ten million stars. While globular clusters spend the bulk of their lives either above or below the plane of the Galaxy, their elongated orbits take them through the disk periodically, and indeed some globular clusters actually spend their lives in more circular orbits within the disk. But the majority are in highly inclined orbits lasting hundreds of millions or years.[1]

These clusters are ancient, having formed when the Galaxy was young and before the disk stars were born. It is known, for example, that many cluster stars are white dwarfs, the old-age relics of Sun-like stars that lived for perhaps ten billion years before reaching their final evolutionary destination. The age of globular clusters also means that the more massive and correspondingly short-lived stars that give rise to pulsars would have lived and died long ago; even if reborn as pulsars they too would have ceased flashing long before the birth of our

Sun. It was thought that globular clusters would be an unlikely place to find a pulsar. That's why, in the 1980s, astronomers were surprised to find that not only were globular clusters home to pulsars, but were in fact among the most prodigious breeding grounds for pulsars anywhere in the Galaxy. Globular cluster pulsars were not the lone pulsars resulting from supernova explosions, however, but rather the rapid and long-lived millisecond pulsars that continue to trace out the dynamics of these most densely populated stellar regions.

Globular Cluster Stars

The majority of stars in globular clusters are highly evolved, low mass stars that were made from the primordial material of the Galaxy. The way astronomers know this is by looking at what they're made of. Stars like the Sun are part of a second generation of stars that contain higher proportions of what astronomers call 'metals'. Now these aren't the same as the metals referred to by chemists; astronomers take a much simpler view of the composition of the Universe. To them, a metal is any element heavier than helium, such as oxygen, iron, silicon and nickel. These elements – the same that make up the terrestrial planets and all living things – were 'cooked' inside the first generation of stars. None of the metals existed before they were made in the first stars. Because there is a lack of metals in the stars of globular clusters, these must be the original stars made of little else than the hydrogen and helium created in the Big Bang. Many of the first generation stars have long since ended their lives, their existence marked only by the slowly fading white dwarfs, and there will be no second generation. There is little hydrogen gas left in these clusters to make new stars since whatever there was originally was long ago used up in making the stars. These are barren clusters: new stars do not shine here. You might think the lack of star formation makes globular clusters rather boring places to visit, dull and monochrome compared with the dynamic and colorful stellar factory that is the Milky Way's disk. Far from it: globular clusters have been found to be immensely productive factories churning out millisecond pulsars faster than anywhere else in the Galaxy. How do they do it, and why?

Near Misses

Clusters of stars are common in the plane of the Milky Way and many can be seen with the naked eye. But these contain young, juvenile stars that have only recently been formed from nebulae. Within a relatively short time 'galactic clusters' dissipate, their brethren stars gently drifting away from each other to wander the Galaxy alone. But globular clusters are different: they are ancient yet the clusters have remained intact.[2] Despite the fact that the closest globular clusters are a hundred times further than the nearest stars in the disk, many are visible to the naked eye.[3] This is because they are made up of so many stars

Figure 11 HST image of the globular cluster 47 Tucanae. This cluster is home to dozens of pulsars. (Courtesy NASA and Ron Gilliland (Space Telescope Science Institute).)

compacted into an extraordinarily small region of space. With hundreds of thousands of stars concentrated in a sphere from 20 to 400 light years across (with 60 light years being typical), they have a density perhaps a thousand times higher than that near the Sun. The sky of a hypothetical planet orbiting one of these stars would be continually ablaze; night would be unknown. Even seen from Earth these clusters are spectacular. Take the cluster omega Centauri. This cluster is so bright it earned a place as an honorary 'star', hence the name. Look at it through a modest telescope, however, and you can see it for what it really is: a spherical collection of more stars than can be imagined, each one crying for attention. It is a wonderful sight.

In the densely populated space within a globular cluster – an environment where a million stars may crowd into a volume with the dimensions of the distance between the Sun and its nearest star, alpha Centauri – close encounters between stars are more than inevitable, they are common. Despite this, actual collisions between stars are surprisingly rare. Just as in the disk of the Galaxy, most of the stars in a globular cluster are in binary systems. If a lone star comes too close to a binary, energy from the motion of the binary is transferred to the lone star, flinging it away. It is chaotic motion like this that keeps the clusters inflated. Rather than collapsing entirely under their mutual gravitational attraction into a single massive black hole, the close encounters between the stars are like swirling dancers linking arms and hurling each other away to more distant reaches of the dance floor. Individual stars rarely leave the dance, however: the gravitational attraction of the cluster as a whole holds a tight reign on its members. Globular clusters remain largely intact.

Not all close encounters between globular cluster stars end so peacefully, however. If two individual stars pass close by one another – say within a few stellar diameters – gravity raises tidal bulges on the surfaces of the stars just as the Moon raises (much smaller!) tidal bulges in the Earth's oceans and even the land itself. These bulges transfer energy of motion to the surface of the stars where oscillations are created, like huge ocean waves. Being robbed of energy of motion, the stars travel more slowly and, if enough energy is transferred, the mutual gravitational attraction between the stars will win out over their velocity, binding them together. The stars form a binary system. At first the orbit is highly elliptical, but in time more and more energy is transferred and the orbit becomes increasingly circular. After a while the stars settle down into a close binary relationship with a distance roughly twice the distance of the initial close approach. For two stars to produce a close binary system like this they have to pass within a mere three times their radius in order for sufficient energy to be robbed from their motion. Within the close confines of a globular cluster, however, such encounters are relatively common. More importantly for our story, a close binary system is an ideal environment for the creation of millisecond pulsars.

X-Ray Sources in Globular Clusters

The first signs that globular clusters would be fertile grounds for the creation of millisecond pulsars emerged in the 1970s when space-borne X-ray telescopes revealed that they contained a high proportion of X-ray sources, perhaps ten times the number of sources found in the Galaxy's disk. The luminosity of the sources and other factors suggested that the X-rays were the result of mass loss in binary systems. Then in the 1980s, two X-ray sources in the globular clusters M15 and NGC 6624 were found to be fluctuating at intervals of 11.4 minutes and 8.54 hours respectively. Such rapid variation implied that they were associated with binary systems with short periods, and that meant close orbits where material was being dragged from one star and dumped on the other. It wasn't long before it was realized that X-ray sources, and hence binary systems, were a hundred times more common in globular clusters than elsewhere in the Galaxy. By the mid-1980s, all-sky surveys for millisecond pulsars had been relatively unsuccessful. As we saw in the last chapter, the connection between X-ray binaries and the creation of millisecond pulsars was already established. Given their dense environments leading to close stellar encounters, globular clusters seemed a logical place to search for millisecond pulsars.

Astronomers, including Andrew Lyne, began searching globular clusters for millisecond pulsars in the 1980s. In 1987 Lyne discovered PSR1821-24, a 3ms period pulsar in the globular cluster M28 in Sagittarius. At first all he had to go on was a radio source with characteristics common to pulsars, but the signal was very weak. After five hours searching the data with a Cray XMP computer, the periodicity of the pulses was found. This was followed a few months later by PSR 1620-26 in the globular cluster M4 in Scorpius. With a period of 11ms, this millisecond pulsar was in an almost circular orbit lasting 191 days, rather a long period for such systems. By the end of the 1980s, it was clear that both Low Mass X-ray Binaries and millisecond pulsars were common in globular clusters, adding further evidence to the theory there was an evolutionary connection.

Just how often two stars will form a binary depends on a large number of factors, including the size of the cluster, its density, and the size of the cluster's core. Generally speaking, however, the original neutron stars that were formed early in the life of a globular cluster from the more massive stars tend to aggregate near the center of the cluster, its most densely populated region. They do this simply because they are so much more massive. Here in the heart of the cluster, stars are on average as close as the Sun and Pluto and so encounters between stars are much more likely. In the ten billion years of the globular cluster's history, many of these fading neutron stars were captured in this way and spun up to millisecond pulsar status. And now astronomers are beginning to find them.

Over a hundred millisecond pulsars in globular clusters have now been found. Ongoing searches for globular cluster pulsars are predominantly being carried out by four observatories: Arecibo (22 globular clusters), the Green Bank Telescope (about 20), Giant Meterwave Radio Telescope in India (10) and, of course, the Parkes Radio Telescope (over 60). By studying globular cluster pulsars, astronomers can learn about the dynamics of clusters. As we've seen, pulsars make excellent motion sensors owing to their exquisitely accurate periods. Globular clusters are spherical and this means millisecond pulsars will sometimes be found on the side of the cluster nearest us, sometimes on the far side of the cluster. By carefully studying and comparing the periods of pulsars, and their dispersion measure, astronomers can learn about not only the motion of stars within globular clusters, but also about the amount of gas within the clusters.

Terzan 5 and 47 Tuc

One of the most heavily pulsar-populated clusters studied so far is Terzan 5, 28,000 light years away in the constellation of Sagittarius. Before 2004, this cluster was known to contain three millisecond pulsars. Using the 100 meter Robert C Byrd Greenbank Radio Telescope in West Virginia, a team of astronomers led by Scott Ransom from the National Radio Astronomy Observatory studied Terzan 5 over a six hour period. This intense search revealed no less than 14 millisecond pulsars, and over the following months discovered a further seven. This raised the total number of millisecond pulsars in Terzan 5 to 24. The latest count at the time of writing was 33. But Terzan 5 isn't the only globular cluster to be scrutinized for pulsars. Another example is the southern cluster 47 Tucanae. Using the Parkes Radio Telescope in the 1990s, Dick Manchester and his colleagues discovered 11 millisecond pulsars in this cluster. When a new instrument called the Parkes Multibeam Receiver[4] was installed it was used to take a closer look at 47 Tucanae and more pulsars emerged. The total now is over 20, and the total number of pulsars in globular clusters now stands at 129 in 24 clusters.

The pulsars in these two clusters are subtly different. While the range of periods for the pulsars in 47 Tucanae range from 2 to 8 milliseconds, those in Terzan 5 span the range from 1.397 to 80 milliseconds. This means there are two distinct populations of pulsars in these two clusters which may be the result of the different conditions in which the pulsars were formed. The most important difference is the density of stars in Terzan 5, which is about twice that of 47 Tucanae. Higher density means more interactions between the stars, which might interfere with the recycling process needed to spin up neutron stars, leaving them spinning slower than they would if they were left alone.

Ever Faster; Ever Stranger

Not surprisingly, such a nurturing environment is bound to give rise to some spectacularly successful millisecond pulsars, and to finish off our tour of globular clusters we will briefly mention two examples. As we saw in the previous chapter, the first millisecond pulsar discovered in 1982 by Don Backer was also the fastest for 24 years. With a period of 1.558 milliseconds, Backer's discovery spins 642 revolutions per second. To put this into perspective, a kitchen blender spins up to 500 times per second. Then in December 2004, a team of astronomers lead by one of Ransom's colleagues, Jason Hessels, discovered a faint pulsar suspect in Terzan 5, confirming its existence a month later. PSR J1748-2446ad turned out to be part of a binary system, orbiting a small (0.14 solar mass) star every 26 hours. The distance between the two stars is about three times the diameter of the Sun. To make matters more difficult, the faint pulsar was eclipsed by its companion each orbit, making it difficult to measure how fast it was rotating. Eventually, however, Hessels and his team were able to more accurately measure the parameters of the pulsar. A year after the discovery the team was convinced they had beaten Backer's record and made their announcement: the Terzan 5 pulsar was spinning 716 times a second, the fastest pulsar known.

Black Widows

Another important discovery about millisecond pulsars in globular clusters is that many of them belong to a class of pulsars called 'Black Widows', so called because they destroy the stars that once breathed life back into their stellar corpses. The original Black Widow Pulsar was discovered in 1988 by Andrew Fruchter. PSR 1957+20 spins every 1.6 ms and is locked in an 8 hour orbit with a white dwarf with a mass only 2% of the Sun. Each orbit, the white dwarf passes between the pulsar and Earth, occulting the signal for almost an hour. What is even more interesting is the fact that a few minutes before the eclipse and for about 20 minutes after, the signal from the pulsar is delayed as it passes through a haze of ionized plasma that completely surrounds the white dwarf. The source of the plasma is the white dwarf itself: it is slowly but surely being eroded by the intense radiation of the millisecond pulsar. Over the next billion years, the pulsar will have completely destroyed the white dwarf to which it owes its existence. Since then, the majority of black widow pulsars have been discovered in globular clusters.

We have now searched the Galaxy and its surrounds for pulsars, and have seen how they are made and where. But this is not the end of the story, for now we can take a brief tour of some of the more exciting individual discoveries that have been made over the last half century since Hewish and Bell's initial discovery. Although necessarily brief, these glimpses will reveal just how much a dead star can tell us about the cosmos.

References

1. By studying the motion of globular clusters, astronomers can trace out the structure of the Milky Way itself.
2. By measuring the motion of globular cluster stars, astronomers can study the cluster's dynamics, including its mass. This has revealed a major puzzle concerning the nature of globular clusters: why they contain so little dark matter. The Milky Way is surrounded by the stuff, and in fact globular clusters drift through this dark matter halo continuously, yet apparently oblivious to its existence. Even more puzzling is the fact that there are other collections of stars similar in size to globular clusters that are saturated with dark matter. The Milky Way is attended to by an entourage of dwarf galaxies whose masses are comparable with the largest of the globular clusters. Dwarf galaxies show a trend of inverse proportion of dark matter with size: the smaller the dwarf, the greater the proportion of dark matter. In defiance of

this, globular clusters contain virtually no dark matter at all. The reasons for this are unclear and the matter is further confused by the fact that the globular clusters and dwarf galaxies formed at around the same time. For some reason, the formation of globular clusters involved only ordinary matter, while dwarf galaxies are dominated by dark matter.

3. The brightest are 47 Tucanae, Omega Centauri in the southern hemisphere, and M5 and M13 in the northern hemisphere.

4. We will look more closely at the contributions of this magnificent instrument to pulsar astronomy in Chapter 15.

12 'Pulsar Planets'

No One was Expecting This...

The announcements were sensational, bordering on the incredible. Within months of each other, two independent teams of astronomers announced the discovery of extrasolar planets, that is planets orbiting stars other than the Sun. If true, the discoveries would be of tremendous importance. Firstly, the discovery of extrasolar planets held deep scientific and philosophical implications, the latest major advance in the Copernican revolution. But what was even more extraordinary about these planets was where they were found. Rather than orbiting a star like the Sun, these planets showed up in the last place anyone expected: orbiting a pulsar, the remains of an exploded star.

One team claimed to have discovered a planet orbiting a normal pulsar in the constellation Scutum; the other announced the existence of two planets and possibly a third orbiting a millisecond pulsar in Virgo. The astronomers making the claims were all experienced and, mindful of the implications, cautious. Theorists were perplexed as to how a planet, let alone *planets*, could come to be in orbit around pulsars, but given the track record of the astronomers making the claims there seemed no reason to doubt the planets' existence. Astronomers around the world launched into serious and intense efforts to explain the pulsar planets. Such cutting edge science sometimes goes awry, however. Within a year, only one claim would prevail: while one discovery would be confirmed as one of the most unexpected and to this day rare finds in astronomy, the other would be revealed as the result of an error. Both would go down in history as prime examples of science at its finest.

The Need for Extrasolar Planets

The importance of the discovery of extrasolar planets cannot be understated. Modern astronomy has been a journey away from the Earth-centered universe. Initially the Copernican revolution was theoretical and insubstantial. Pre-telescopic observations reached their peak with Tycho Brahe and their theoretical interpretation by Johannes Kepler. Galileo, the first to make major astronomical discoveries, used a telescope that is today tiny by even amateur standards. The list of his discoveries is well-known: the phases of Venus, the craters on the Moon, the satellites of Jupiter, and so on. These and other discoveries shocked the world out of its geocentricity; at last it became clear that

Earth was a planet orbiting the Sun, then the Sun took its place as one of the myriad stars. As astronomical telescopes and detectors improved and extended into other bands of the electromagnetic spectrum, the Milky Way became but one of billions of galaxies in the observable universe.

Of course one of the ultimate questions was (and still is!): is there life out there? Some have paid for such suggestions with their lives, but the question itself is undying. The stars are other suns, but in order to have life you needed to have planets for it to live on. Earth-like planets capable of supporting life, perhaps even intelligent life. The enormity of such a discovery is detailed in many books, and indeed the search for extraterrestrial intelligence, or SETI, is now an important scientific discipline. What was needed, of course, was evidence that extrasolar planets existed at all. In their 1966 book *Intelligent Life in the Universe*, I. S. Shklovskii and Carl Sagan point out that:

> 'The existence of dark companions of the nearest stars, the argument from stellar rotation, and the contemporary theories of the origin of solar systems together strongly point to a plurality of habitable worlds. But only future developments in astronomy can demonstrate beyond the shadow of a doubt the existence of large numbers of such planets.'

A quarter of a century later the frustrating of lack of evidence for planets around other stars persisted, although there was an accumulating number of tantalizing clues. In the 1980s evidence of embryonic solar systems began to emerge with discoveries of disks of dust around young stars such as β Pictoris. Other stars showed flows of gas from their poles traveling at hundreds of kilometers per second. It was known that hot winds blow from young stars, but what made the winds blow in opposite directions of a newly formed star must be a constraining disk of dust and debris surrounding the star's equator which, presumably, formed at the same time as the star. But a dust disk is not a planet. Astronomers could only be satisfied with the detection of a planet, but this was and still is an enormous technological challenge.

How to Find Extrasolar Planets

The difficulty in detecting a planet around another star is the incredible faintness of planets compared with stars. Take our Solar System, for example. Compared with the brilliance of the Sun even the largest planet Jupiter is almost a billion times fainter at visible wavelengths and several thousand times fainter in the infrared. Even if it was possible for a detector to span such a tremendous range in light intensity, the fact remains that seen from even a modest distance of 30 light years, Jupiter would be a mere half second of arc away from the Sun. Even a giant planet like Jupiter, five times farther from the Sun than Earth and eleven times Earth's diameter, is lost in the solar glare. Searching for even the largest extrasolar planets seemed hopeless. There had to be another way.

A Dynamical Approach

As we've seen, when two stars or pulsars share an orbit they influence each other gravitationally, and the same is true for planets orbiting a star. While we say planets orbit stars, it's more accurate to say a star and a planet share a common center of mass. The center of mass is not in the middle of the star but slightly closer to the star's surface in the direction of the planet. This causes the star to rotate around an offset center of mass as the planet revolves around it. The effect is tiny, but theoretically detectable. First, the star would appear to move from side-to-side as it traveled across the sky. Careful astrometry revealed the existence of the white dwarf companion to Sirius, Sirius B, because of this effect. Could planets be detected in the same way? One of the first claims for the astrometric discovery of an extrasolar planet emerged as early as the 1960s. Based on 24 years of observations made using the 24' telescope at Sproul Observatory, Peter van der Kamp announced that two planets roughly the size of Jupiter were orbiting Barnard's star. Unfortunately follow-up observations at other observatories showed that the planets did not exist. Alternatively, as the star moves towards and away from us the star's light will be alternately blue shifted then red shifted. But with a change to the velocity of a star caused by an orbiting planet likely to be tiny, perhaps only a meter per second, it seemed unlikely that such measurements were bearing fruit any time soon.

A Pulsar Planet in Scutum

The difficulties involved in detecting extrasolar planets were summarized in the opening remarks of an extraordinary paper published in July 1991 by Andrew Lyne, Matthew Bailes and Setnam Shemar. The paper reported observations of the pulsar PSR1829-10, a normal pulsar 30,000 light years away in the constellation Scutum near the center of the Galaxy. What was unusual, however, was that this pulsar seemed to be orbited by a planet.

PSR1829-10 was discovered along with about forty other pulsars in 1985 using the 76 meter Lovell radio telescope at Jodrell Bank. Since its discovery, Lyne had kept an eye on PSR1829-10. Like other pulsars, PSR1829-10 was slowing down, and the rate of slow down suggested it was about a million years old. But what intrigued Lyne was the way in which the pulsar was slowing: rather than a steady decline, the period of the pulsar seemed to slow and speed up rhythmically with a period of about six months. The changes were small, just 0.008 seconds, but they wouldn't go away. Other pulsars had shown similar variations in arrival times, and these had indicated the existence of companion stars. What was unusual about Lyne's discovery was the mass of the companion implied by these periodic variations: just ten times the mass of Earth, or a thirtieth the mass of Jupiter. This was too small to be a star. It was close to the pulsar, too: following an almost circular orbit with a radius of 0.7 AU, the object was about as far from the pulsar as Venus is from the Sun. The only logical conclusion was that Lyne,

Bailes and Shemar had discovered a body less than the mass of the smallest star orbiting PSR1829-10. They had found the first extrasolar planet.

The news was sensational; the theorists were perplexed. Aside from the importance of the announcement of an extrasolar planet, the fact was that there shouldn't be planets orbiting a pulsar. After all, pulsars are the ashes of supernovae, and there seemed little doubt that, even had the original star had a family of planets, once it went supernova the planets would all be destroyed. At the very least, during a supernova the star would lose so much mass that it would release its gravitational grip on any attending planets, releasing them into deep space. Alternative suggestions were made for the perturbations. One was precession of the pulsar's axis making it wobble like a top. Lyne pointed out that this would also affect the pulsar's radio emissions, and there was no sign of this. Another explanation was that there could be changes occurring within the pulsar itself as the liquid interior of the pulsar moved. Although this phenomenon had been observed in the Crab pulsar, the fact remained that it was an irregular effect and was not seen in older pulsars like PSR1829-10. It seemed that, in the most unexpected place, the first planets outside our Solar System had at last been found.

The astronomers published their results in the July 26th edition of the prestigious journal *Nature*. The announcement set the theorists to work. How could a planet possibly exist in such an inhospitable environment? Also, this wasn't the first time someone had proposed the idea of a planet orbiting a pulsar. In the late 1970s, two Polish astronomers suggested that one explanation for the behavior of PSR 0329+54 could be explained by the existence of a planet up to about half the mass of Earth. In the years since, the planet was not confirmed. Aside from the reality of such a pulsar planet, there seemed no way a pre-existing planet could survive the supernova explosion, and so it must have come along later. One possibility was that the planet was formed as a result of the supernova itself. If some of the material thrown off during the explosion somehow found its way back to the pulsar, it could conceivably accrete into a planet just as protoplanetary material does around normal, young stars. Alternatively, the pulsar planet may have formed from a disk of material deposited by a companion star as it was torn apart by the neutron star. Yet another explanation was that the pulsar was formed from the merger of two white dwarf stars, and that during the merger a disk of material was formed around the pulsar, which later accreted into a planet. Or perhaps it was the aftermath of the collision or at least close encounter between a pulsar and a normal star with an attendant family of planets. Although there was still no consensus on how the planet came to be, after some initial objections it now seemed that, as unexpected as a pulsar planet was, its existence wasn't as improbable as first thought.

Wolszczan and Frail's Discovery

Two years before the announcement of PSR1829-10, the 300 meter Arecibo telescope was undergoing repairs to cracks in its structure. Although the huge

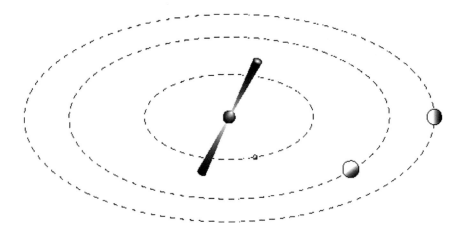

Figure 12 Planets orbiting around the millisecond pulsar PSR B1257+12 (not to scale) discovered by Dale Frail and Alex Wolszczan in 1991. Illustration by the author.

instrument could no longer be maneuvered, it was still able to receive signals from space, albeit limited to whatever patch of sky passed over head through its beam. The usual long queue of astronomers suddenly dwindled, allowing a research associate Alex Wolszczan, working for the National Astronomy and Ionospheric Center at Arecibo to use the dish for extended periods of time. His goal was to search for millisecond pulsars away from the plane of the Galaxy. While observing from Arecibo, Wolszczan discovered a pulsar subsequently called PSR1257+12 about 1,200 light years away in Virgo. But a more attractive discovery was a binary pulsar similar to the one Joe Taylor and Russell Hulse had discovered. Wolszczan continued to monitor PSR1257+12, but it took a back seat as Wolszczan concentrated on the binary.

It wasn't long, however, before PSR1257+12 grabbed Wolszczan's attention: the pulsar was behaving strangely. Like PSR1829-10, Wolszczan's pulsar was not spinning regularly. Now this pulsar was an example of a millisecond pulsar, and these stars are far more stable and reliable than their slower counterparts. Further, while the perturbations seemed to be regular, the data available was frustratingly incomplete. Importantly, the exact position of the millisecond pulsar was not accurately known. Because pulsars are observed from a planet in orbit around the Sun, the exact arrival time of the individual pulses depends on the Earth's position in its orbit, which determines whether the Earth is approaching or receding from the pulsar. This motion causes a Doppler shift of varying amounts, so the position of the pulsar in the sky has to be known as accurately as possible in order to properly compensate for the Earth's motion.

Wolszczan contacted Dale Frail, an astronomer working at the National Radio Astronomy Observatory's Very Large Array in New Mexico. The VLA, which employs the interferometric technique discussed in Chapter 3, consists of 27 radio dishes mounted on rails forming a giant Y. It should have been possible to pinpoint the location of PSR1257+12 using the VLA, but at first Frail, who only had limited access to the telescope, had difficulty narrowing down the pulsar's position. Eventually, however, Frail was successful. It was during this period of collaboration that the news came from across the Atlantic that Lyne, Bailes and Shemar had discovered their pulsar planet.

With the discovery very much on their minds, Wolszczan and Frail continued their investigation of PSR1257+12. Frail later recalled that he joked with Wolszczan: 'Don't find any planets.' Eventually, however, there was no escaping the conclusion over what was causing the perturbations in the pulsar's period. Wolszczan sent back an email after having analyzed the data: two planets were orbiting the pulsar.[1] One planet was 3.4 times the mass of Earth and an orbit of 66.6 days; the other was slightly smaller at 2.8 Earth masses and an orbit of 98.2 days. By September 1991 Wolszczan was back at Cornell University but continued receiving data on the pulsar from Arecibo. It was crucial that Wolszczan was able to show that the observations accurately matched the predictions made by the mathematical model of the pulsar planet system. He entered the data and it worked ... sort of. Soon, however, he realized he had assumed circular orbits for the planets, whereas most orbits are elliptical. After three hours of fiddling with the parameters, Wolszczan ran the computer program again and the errors disappeared. He was now satisfied: he and Frail had discovered two planets orbiting the millisecond pulsar PSR 1257+12.

The result was published in the January 9th 1992 edition of the journal *Nature*, almost six months after Bailes, Lyne and Shemar had published their discovery in the same journal. That month, Frail and Wolszczan set off for Atlanta, Georgia for a meeting of the American Astronomical Society where they intended to present their findings to the astronomical community. Also presenting at the meeting was Andrew Lyne, but what he was to present was set to both shock and impress everyone present.

Error

Like Wolszczan and Frail, Lyne also had to take into account the arrival times of the pulses, which depended on the Earth's motion, which in turn depended on the position of the pulsar. And like Wolszczan, Lyne did not know the position of PSR1829-10 very well to start with. The corrections were on the order of 500 seconds, the time it takes light to travel the distance between Earth and the Sun, and the corrections have to be accurate to within a few microseconds. This requires high precision and involves a long and complicated calculation to get it right. It was some months after the publication of the paper in *Nature* that Lyne had some quiet time between Christmas and the New Year and went into the deserted observatory to do some more work on the system. For some reason, he decided to reprocess all the data he had on the pulsar from the start. If there was an error in the position, an annual sinusoid should appear in the data, and from the amplitude and phase of this it was possible to get a better estimate of the position of the pulsar. Lyne and his colleagues used a simple differential correction model and a fitting program to remove the annual sinusoid and determine a better position. Unknown at the time, the software for this differential correction was flawed and did not take account of the small eccentricity of the Earth's orbit. Ordinarily, this wouldn't have mattered since it was their usual procedure to recalculate the corrections for the new position and repeat the analysis. In the case of PSR1829-10, for various reasons, this procedure was not followed. The result of the improper differential correction was a 6-month sinusoid in the data which they had attributed to the existence of a planet. The error in the correction was only about 1 part in 100,000, but it gave a 100 standard-deviation planetary signature. When Lyne reanalyzed the data, there was no planetary signature. After the sensation of the announcement and the subsequent theorizing over the existence of a pulsar planet, it turned out that PSR1829-10 was in fact a solitary pulsar. There was no planet there after all.

It took Lyne only five minutes to figure out the cause of the error. Then he just sat there frozen to his chair for the next half hour as the enormity of the mistake sunk in. This had happened in spite of all the care the astronomers had taken, all the checks they had made on other nearby pulsars. Yet this was more than a scientific error: many people had been enthralled by the discovery and hundreds of astronomers had spent months trying to interpret it. Nonetheless, there was only one thing to be done: put the record straight. He could have simply published a retraction, but in a display of extraordinary scientific courage and honesty, Lyne decided to announce the error at the American Astronomical Society meeting, the same one that Wolszczan and Frail were attending.

Lyne began his address with, 'This talk is not the one I was originally proposing to give.' As he explained the error in front of a thousand of his colleagues, he appeared shaken, and there was a gasp from the audience as they comprehended what he was telling them. There was no planet around PSR 1829-10. This was a dreadful period for Lyne. 'It was an indescribable embarrassment, and resulted from a major failing of my scientific methodology,' he told me. The astronomical community seems to have concluded otherwise. Acknowledging his scientific honesty, and the fact that the same thing could have happened to anyone in the hall, the audience erupted into loud, sustained applause. Lyne says that he was amazed at the understanding and sympathy that many experienced astronomers showed. 'But the chilling thing is that I do know that I could still make a similar mistake again,' he continued. 'When stretching technology and knowledge to the limit, we often have to take short-cuts and make economies. We try to anticipate the effects of these, but there can always be little crevices in the fabric of a complex experiment in which a little demon can lurk, waiting to trick you...'

Wolszczan's Turn

Lyne's announcement made things even more difficult for the next speaker: Alex Wolszczan. He knew he had to be even more convincing than he had planned, but he was confident in his analysis and results. Wolszczan had been meticulous in ruling out possible errors, and he accounted for these in his presentation. It was clear this time that there was no mistake: there were indeed planets orbiting around PSR B1257+12. Two years later he published a second paper in the journal *Science* offering proof of the planets' existence. Not only were the planets influencing the rotation of the pulsar, leading to their initial discovery, but they were also influencing each other. By carefully analysing their mutual gravitational tugging, Wolszczan was able to reveal the existence of a third planet, this time about the size of the Moon and orbiting the pulsar every 25.3 days. What's more, the planets seemed to be orbiting the pulsar in the same plane, suggesting a common origin rather than a post-supernova capture of the planets. Independent confirmation of the planets around PSR B1257+12 was made by Shauna Sallmen and Roger Foster, an important step considering the history of pulsar planets.

The following year, Don Backer and his colleagues carried out an analysis of a pulsar in the globular cluster M4 in the constellation Scorpius. This pulsar was discovered in 1988 by Lyne and his colleagues and was only the fifth millisecond pulsar found and the second pulsar in a globular cluster. Lyne initially reported the pulsar as being part of a binary, sharing a common orbit with a low mass companion. Backer realized, however, that in order to fully explain the observed variations in the pulsar's period a third object was needed. At first it was assumed that the third object was also a star. Follow up work by a number of scientists differed in their estimates of the mass of the third object, but all agreed that it was too small to be a star. In 1993 Stephen Thorsett, Joe Taylor and Zaven Arzoumanian announced that there were two possibilities for this system. Either the binary pulsar was orbited by a third star at a distance of 50 AU, or the pulsar-white dwarf pair were being orbited by a planet perhaps five times the mass of Jupiter. Either way, they pointed out, this was an unusual system: whether it was the first pulsar triple system, or the first planet orbiting a binary pulsar, PSR B1620-26 was an unusual find. As it turned out, the planet hypothesis was best supported, and PSR B1620-26 remains only the second pulsar planetary system confirmed.

At last the existence of extrasolar planets around pulsars had been demonstrated and although the explanation for their existence was still being debated, it had the flow-on effect of providing a push for other searches for extrasolar planets. Over the following years, dozens of planets would be discovered around other stars using a variety of detection techniques. With each technological advance, the size of the planets has become smaller, and it is anticipated that one day planets the size of Earth will be discovered orbiting Sun-like stars with the ultimate aim of finding a planet that supports life. Ironically, the first planets to be discovered are barren: bathed in the intense radiation of the pulsar, they are mere rocks hurtling around the pulsar.

Whether or not pulsar planets are unusual is still unknown, but in 2006 Deepto Chakrabarty of MIT announced the discovery of a disk of debris surrounding a pulsar called 4U 0142+61. If this disk turned out to be real, it could be evidence that the formation of pulsar planets is common and many more await discovery. But 4U 0142+61 is no ordinary pulsar: with properties that make ordinary pulsars look positively tame, this new breed has opened yet another chapter in the story of pulsar astronomy: magnetars.

Reference

1.   DuBois, Charles 'Planets from the start', Research/Penn State, Vol. 18, No. 3, September 1997.

13 'Magnetars'

Introduction: A New Breed

Most scientists found the idea too much to swallow. In 1992, Robert Duncan and Christopher Thompson put forward a model of a new type of pulsar, a pulsar with extraordinary properties. The theory attempted to explain a number of strange, so far unexplained sources of gamma rays called Soft Gamma Ray Repeaters (SGRs). These distant objects unpredictably emitted strong bursts of gamma rays, quietening down over a matter of seconds, and then remaining dormant for years only to flare up again unexpectedly. What was curious about these objects was the same thing that attracted Jocelyn Bell's attention more than two decades before. Just as Bell's 'scruff' turned out to be a periodic signal, so the powerful if brief emission of gamma rays from SGRs was not a steady hiss, but a series of spikes. This implied rotation of a small, energetic, and dense object, and the best candidate available was a neutron star. What Duncan and Thompson proposed was that SGRs were a new breed of pulsar, perhaps even outnumbering the normal ones. This new type of pulsar had an important distinguishing characteristic: its magnetic field was hundreds, perhaps thousands of times more powerful than those known to be powering ordinary pulsars. Like so many new horizons of astronomy, the discovery of SGRs that led to the discovery of these magnetic monsters came about completely by accident.

Gamma Rays from Space

We saw in Chapter 6 that gamma ray astronomy emerged in the mid-1960s, and that from the very beginning both gamma and X-ray astronomy was to be instrumental in understanding pulsars. Gamma rays, you'll recall, are a form of electromagnetic radiation just like light and radio waves; the difference is they pack more energy per photon, even more than X-rays. In fact, having extremely short wavelengths of less than 10^{-5} meters, gamma rays are the most energetic and the most penetrating of all the forms of electromagnetic radiation available to astronomers. Ironically, neither gamma rays nor X-rays can penetrate the Earth's atmosphere. Only in the early 1960s when satellites equipped with gamma ray detectors were launched, were gamma rays from space detected, and a new branch of astronomy emerged. It was this development that led to Gerry Fishman's discovery of the Crab pulsar signal, detected prior to Bell's discovery. This isn't to say high energy astronomical phenomena weren't anticipated.

Theoretical work in the first half of the twentieth century showed that violent astrophysical phenomena should produce high energy photons, that is, gamma rays, and so in 1961 a spacecraft called Explorer XI was launched carrying a gamma ray detector. The satellite detected less than a hundred gamma ray photons which appeared to be coming from random directions in the sky. As often happens in astronomy, however, some of the most fruitful discoveries came from an entirely unexpected source.

Astronomical Fallout

In the 1960s, a treaty was signed by the nuclear nations of the time not to detonate nuclear bombs for testing purposes either in the Earth's atmosphere or in space. In 1963 the US Air Force launched a series of satellites called Vela designed to monitor for such nuclear explosions in the atmospheric and space environments. Operating in pairs, the satellites kept vigil for gamma ray emissions that would signify a nuclear detonation. Two satellites would detect the gamma rays at different arrival times allowing scientists to triangulate and plot the location of the source. But in 1965 the satellites began detecting bursts of gamma rays that were clearly not of terrestrial origin. The events lasted anywhere from a tenth of a second to a thousand seconds. Although not of immediate importance, the records were carefully filed away by Ray Klebesadel of Los Alamos Scientific Laboratory for later study. There they remained until 1972 when they were analyzed by Ian Strong, also of Los Alamos Scientific Laboratory. Strong, Klebesadel and Roy Olsen found that the accuracy of the observations was sufficient to eliminate sources within the Solar System; in fact they seemed to be coming from random directions in the sky.[1] The following year they published their results, showing that gamma ray events were coming from cosmic sources. These sudden, unexpected bursts of gamma radiation naturally became known as gamma ray bursters, or GRBs.

5th March 1979

Following the discovery of cosmic gamma rays a number of satellites were launched to study them, including the US Small Astronomy Satellite, SAS 2 in November 1972, and the European Space Agency's COS-B in August 1975. These and other spacecraft revealed a violent face of the Universe but, despite identifying individual sources of gamma rays, the observations lacked the resolution to link these sources with objects visible at other wavelengths. By 1979 hundreds of events had been detected, and technology was becoming ever more sophisticated. Spacecraft were roaming the Solar System, some looking into deep space, some at the Sun, still others exploring the planets. Most were not looking specifically for cosmic gamma ray sources. Then, on March 5th that year, an unprecedentedly intense burst of gamma rays swept through the Solar

System. The burst triggered detectors aboard no less than ten spacecraft. The first to notice something big was happening was a pair of Russian spacecraft Venera 11 and Venera 12. In December 1978, the spacecraft had flown past Venus delivering landers to the surface of the planet before gliding into orbits around the Sun. As they drifted through the inner Solar System they recorded the background level of gamma radiation. Suddenly the gamma ray detectors on board each of the spacecraft recorded a burst of gamma radiation a hundred of times the normal flux in a fraction of a millisecond. Having triggered the twin Venera spacecraft detectors, the wave sped on through the Solar System at the speed of light. Next to detect the wave was NASA's Helios 2, then the sensors aboard Pioneer Venus Orbiter, which was at the time scrutinizing our sister world. Then it reached Earth. The wave of gamma radiation swamped the detectors of three Vela satellites, the Soviet Prognoz satellite, and the Einstein Observatory, a satellite designed to explore the high energy universe. The wave continued out through the Solar System, but before it left it washed over the International Sun-Earth Explorer spacecraft. One after another, the gamma-radiation wave sparked the attention of these robot explorers.

The event was as powerful as it was unexpected: a hundred times stronger than anything that had been previously recorded. Over the following three minutes the signal gradually faded. Fourteen and a half hours later a much gentler burst of X-rays emerged from the same source, which continued to emit bursts of gamma rays over the years to come. Assuming this was yet another, albeit more energetic example of what had been seen before, astronomers listed the March 5th event as a gamma ray burster. But aside from the intensity of the gamma ray burst, and the fact that it emitted gamma ray bursts more than once, there was something unusual about this event. When the fading signal was analyzed, it was clear that it wasn't a steady signal, but one that fluctuated with a period of eight seconds.

1982: Location Confirmed as N49 in LMC

Years later, a team of astronomers led by Thomas Lytton Cline used the arrival time of the burst recorded by the individual spacecraft to triangulate the position of the gamma ray source: it was located at Right Ascension 5 hours 26 minutes, Declination –66 degrees, giving rise to its ultimate designation SGR 0526-66. What they found was that the source coincided with the remains of a supernova that had exploded 5,000 years earlier called N49. This matched the idea that gamma rays were produced by violent astrophysical events, but what really electrified astronomers was where N49 was. As powerful as the gamma rays were, astronomers originally assumed that the source was in the Milky Way galaxy, relatively close to the Sun. But N49 was not in the solar neighborhood, in fact it wasn't even in the Milky Way. It lay 180,000 light years away in the Large Magellanic Cloud, a thousand times further than anyone suspected. Could it be a coincidence that SGR 0526-66 was in the same line of sight to N49? This was extremely unlikely given the precise alignment needed between two unrelated objects in space so far removed. The logical conclusion was that the gamma ray source and N49 were one and the same. The implication was astounding: to be detectable at such a distance SGR 0526-66 would have to be a million times more powerful than theory suggested it could be. In a fraction of a second, SGR 0526-66 had put out the same amount of energy as the Sun emits in 10,000 years.

It was clear that whatever was producing the gamma rays was not an ordinary star, and that meant one of two candidate sources: a black hole or a neutron star. The eight second period seen in the data ruled out the black hole since this kind of object is rotationally symmetrical and offers no way of producing a regular signal. That left neutron stars. Here was a paradox. The identification of the source with such a young supernova remnant (SNR) implied that the neutron star would be young and energetic, rotating at high speed. Such a sluggish period as demonstrated by the gamma rays implied an aged neutron star that had lost its rotational energy over millions of years. There was still one more piece of the puzzle to emerge from the March 5th event. The source of the gamma rays was close to, but not exactly in the center of, the SNR. If the source of the gamma rays, whatever it was, was produced by the supernova, it had apparently been kicked out of the center of the debris cloud, the supernova remnant. This is not uncommon, with a number of neutron stars known to be far removed from the site of their birth.

More SGR's, More Surprises

As powerful as the March 5th event was, it wasn't in fact the first. Earlier that year, the soft gamma ray repeater SGR1806-20 was detected in the constellation of Sagittarius. Lying in the direction of the galactic center, the true distance to this object is unknown, but may be far more than the 25,000 light years to the heart of the Galaxy. Subsequent observations with the X-ray satellite ASCA showed that SGR 1806-20 was also a discrete source of X-rays. Then a mere nine days after the March 5th event, a third SGR was discovered in Aquila, SGR 1900+14. Over two days, SGR 1900+14 gave off three bursts of gamma rays, then another three in 1992.[2] In a single month, three SGRs had been discovered. In mid-June 1998, an instrument called BATSE[3] aboard the Compton Gamma Ray Observatory, the second of NASAs Great Observatories (the first was the famous Hubble Space Telescope) detected a fourth source, SGR 1627-41. These remain the only known SGRs. However, there were other objects out there that are now believed to be related to SGRs, and which also represent the new breed of highly magnetized neutron stars. In the 1970s, an X-ray observatory called Uhuru (the original Small Astronomy Satellite, SAS-1) was launched to study the X-ray universe. One of the most important discoveries was pulsing sources of X-rays with periods between 6 and 12 seconds. Once again, the high energy of the emissions strongly suggested a pulsar, and it was thought that the X-rays were produced when matter from a companion star spilt over onto the pulsar. Only seven of these strange objects have been discovered so far, and they have become known as anomalous X-ray pulsars, or AXPs.

Less than a dozen objects, not very often seen and consequently not very well studied. But there was enough data to support a radical theory that would introduce a new breed of pulsar into the menagerie. In 1992 Robert Duncan and Christopher Thompson put forward their explanation of both SGRs. Five years earlier they had calculated the strengths of magnetic fields that could lace through newly born pulsars. The figures they came up with, however, were enormous – a thousand times that of known pulsars – and neither really knew what to make of it. Then in 1992 and 1995 they made the connection between their theoretical predictions and the handful of objects known as SGRs and AXPs.[4]

Birth of a New Breed of Pulsar

We've seen how pulsars emerge from the death throes of massive stars: as the core of the star collapses, angular momentum is conserved spinning the pulsar faster and faster. In most cases, the newly born pulsar will be spinning at most tens of times a second. The intense magnetic field surrounding the pulsar slows the rotation, and in time all pulsars spin down and eventually go quiet, disappearing from view. So well understood is this phenomenon that the age of a pulsar can be determined from how quickly it is slowing down: the faster the deceleration, the younger the pulsar. Now imagine not just a massive star, but one so immense that, at the birth of the pulsar, it carried so much angular momentum that the pulsar was spun up to hundreds or even a thousand revolutions per second. Such a fast rotation had an unexpected effect. At the moment of creation the temperature inside the pulsar would be greater than 10 billion K, and at this temperature the fluid interior of the pulsar would be circulating furiously. This convection produces a dynamo effect that transfers about a tenth of the kinetic energy into generating an intense magnetic field which threads through and around the pulsar. Ordinary pulsars gain their energy from rotation and their surrounding magnetic fields of 10^{12} gauss[5] channel energy into the beams we detect here on Earth. The beams carry away energy, causing the pulsar to gradually slow down. This new breed was different. With field strengths approaching 10^{15} gauss, these were the most highly magnetized objects in the Universe, and so Duncan and Thompson gave them a new name: *magnetars*.

Duncan and Thompson's calculations suggested that such an intense magnetic field acts as a kind of brake on the rotation of the magnetar, slowing it from its thousand revolutions a second to one rotation every eight seconds, all within a few thousand years. The coincidence between this prediction and the 6 to 12 second periods of SGRs and AXPs, not to mention the latter's association with young supernovae and neutron stars, was too much to ignore. Further, as the magnetic field continues to slow the magnetar, it becomes increasingly faint

Source	Strength
Earth's magnetic field. Detectable with a compass.	0.6 Gauss
Hand-held magnets, including fridge magnets.	100 Gauss
Sunspots.	4000 Gauss
Strongest sustained magnetic fields created in the laboratory.	4.5×10^5 Gauss
Strongest magnetic field ever created. These are only very short lived.	10^7 Gauss
Strongest fields on non-neutron stars. These are found on only a few strongly-magnetized, compact white dwarf stars.	10^8 Gauss
'Normal' pulsars.	10^{12}-10^{13} Gauss
Magnetars, soft gamma repeaters and anomalous X-ray pulsars.	10^{14}-10^{15} Gauss

Figure 13 Table showing relative strengths of familiar and unfamiliar magnetic fields.

and within a brief period of time the once powerful and spectacular magnetar becomes invisible. Statistically, therefore, magnetars are visible only briefly and would explain why so few SGRs and AXPs had been detected. Although Duncan and Thompson didn't set out to explain SGRs or AXPs, the connection soon made sense.

But what was causing the bursts of gamma rays? Once the magnetic field exists, it's hard to get rid of. As the magnetar rotates the field gets wound up in a spiral, and is continually shifting to less strained patterns. The surface of a neutron star is a solid crust about a kilometer thick, the only stars with a solid surface. The shifting magnetic field continually bends and stretches the crust of the neutron star and causes kilometer long cracks to appear. As the field above the crack thrashes into a new position it releases magnetic energy creating a cloud of electrons and positrons to be discharged along the field lines giving rise to the gamma ray and X-ray emissions. The star's magnetic field releases an enormous amount of energy within a few tenths of a second as it snaps into a new configuration; this is observed as a brief, extremely intense spike of hard gamma rays. Most of the flare's energy comes out in this initial spike, leaving behind an X-ray-bright cloud of electrons and positrons trapped by magnetic field lines near the star, which radiates away its energy within a few minutes, the duration of a giant flare from an SGR.

Confirmation Needed

As much as the idea made sense to them, Duncan and Thompson's theory was not well received. But in October 1993 Chryssa Kouveliotou used the Compton Gamma Ray Observatory to record an outburst of gamma rays. Although CGRO was unable to pinpoint the source of the emissions, the Japanese/NASA Advanced Satellite for Cosmology and Astrophysics (ASCA) turned quickly toward the burster after receiving Kouveliotou's alert, and immediately found a suspicious X-ray point-source within the CGRO error box. This X-ray star gave off another burst of soft gamma rays as ASCA watched, confirming that it was the SGR which astronomers had named SGR 1806-20. For the first time astronomers knew the precise sky position of an SGR in our Galaxy. Then, in 1995, Kouveliotou used NASA's Rossi X-ray Timing Explorer satellite to study SGR 1806-20 in more detail. One of the advantages of RXTE is that rather than just detecting gamma rays and X-rays, it counts them with high temporal resolution and so is ideal for producing data that can reveal periodic signals. The RXTE data revealed SGR 1806-20 had a period of 7.47 seconds. Not only that, but the pulses were slowing rapidly. Now there was little doubt that SGR 1806-20 was associated with a pulsar, and that something was slowing it down at an unprecedented rate. The connection with magnetars now seemed certain.

What a Blast

Until this time, none of the observed events had exceeded the March 5th 1979 event. Then on August 27th 1998 it happened: a giant gamma ray burst swept across the Solar System, and this time it was even more powerful than the one seen in 1979. Instruments aboard seven spacecraft recorded the event, their gamma ray detectors driven to their maximum readings or off the scale. So powerful was the burst that it forced one of the spacecraft, the Comet Rendezvous Asteroid Flyby, into a protective shut down. As the burst swept over the Earth, it pounded the ionosphere causing it to plummet from its normal altitude of 85 kilometers down to 60 kilometers above the surface. As powerful as this burst was, intrinsically it was less powerful than the 1979 event. What made it so powerful as seen from Earth was its distance from us: the gamma rays came from an SGR 20,000 light years away (compared with the 170,000 light years to N45 in the Large Magellanic Cloud). Observations of the last few hundred seconds of the burst made by Kouveliotou revealed a period of 5.16 seconds, and it was slowing down in a similar fashion to SGR 1806-20.

This outburst remained the brightest seen until December 27th 2004. SGR 1806-20 flared again, only this time the emission of energy was enormous even by SGR standards. By then it was known that the SGR 1806-20 was on the other side of the galactic center, some 50,000 light years away and behind dense clouds of interstellar gas and dust. Despite the distance, despite the obscuring clouds, the eruption unleashed more energy on Earth than it receives from a powerful solar flare. It seems the flare released a sizeable fraction of the total energy stored in the magnetar's magnetic field. The event was recorded by at least 15 spacecraft on various missions swamping their detectors. So powerful was the blast that it released in a fraction of a second more energy than the Sun releases in 250,000 years, a hundred times more powerful than anything detected previously and 10,000 times brighter than SGR 1806-20 had shone before. If the SGR had been within 10 light years of Earth, it would have destroyed the ozone layer. Despite the fact that the blast hit Earth on its day side and hence competing with the

Sun, its ionizing effects on the Earth's atmosphere were detectable in amateur instruments. Although the gamma ray event was over in minutes, radio astronomers were able to monitor the afterglow for weeks. Astronomers using radio telescopes in the United States, Australia, The Netherlands and India determined that the magnetar had emitted a non-spherical fireball of radio-emitting material swelling outward at about a third of the speed of light. Other astronomers studying the magnetar in X-rays have managed to probe the structure of the star by examining vibrations set up at the time of the giant flare in a manner analogous to seismologists studying the Earth's interior by recording earthquakes.

More than a Vindication

Observations of the dozen or so magnetars discovered so far continue, with one of the most recent developments being the optical detection of an AXP. Although the amount of light is tiny, it too fluctuates in time with the X-ray pulsations. In its own way, this is further confirmation of the magnetar model. If the AXP had been an ordinary neutron star with an accretion disk – that is, the X-rays produced by matter falling onto the neutron star – then the object would not only be much brighter at visible and infrared wavelengths, it would also not be fluctuating at the rate observed.

The confirmation of magnetars is more than a vindication of Duncan and Thompson's theory. It also raises the possibility of an entire population of so-far invisible magnetars, perhaps hundreds of millions of them. Despite their brief, spectacular displays, the fact remains they are potentially visible for only a brief period of galactic time. It remains with the potential of astronomical instrumentation to detect such stars that are so far beyond our vision.

References

1. Subsequent observations allowed astronomers to not only detect gamma ray bursts, but to plot them on the sky. It is now known that these sources of gamma rays are distributed evenly across the sky, confirming their extragalactic origins.
2. SGR 1900+14 was initially identified with a young supernova remnant, this time within the Galaxy. During the 1990s it was thought that, like SGR 0526-66 in the LMC, SGR 1900+ 14 had been propelled out of the center of its remnant at the time of the explosion. However, Rob Duncan notes that in 2008 'there are now good reasons to believe that the young supernova remnant (SNR) within which SGR 1900+14's sky position resides (displaced from the center) is just a chance overlap of objects at different distances from Earth and that the SGR was not born in the supernova which made that SNR. Careful observations have shown that SGR 1900+14's angular position also

lies within a compact cluster of young stars in the distant galactic disk. Although intrinsically very luminous, the massive young stars of this cluster are not easy to see in optical light; they are bright only in the infrared (IR) because of many kiloparsecs of intervening galactic dust. Such young, compact star clusters cover only a very tiny fraction of the sky in the galactic plane, a circumstance which makes a chance overlap at different distances from Earth unlikely. But the circumstance which really strongly bolsters this interpretation (i.e., that the SGR is actually in the cluster, and not associated with SNR) is that SGR 1806-20's angular position also overlaps with a very rare, very compact cluster of bright young, massive stars far across the Galaxy. The *a posteori* chance of the two most active galactic SGRs having such positions overlapping with these tiny clusters, yet with no physical cluster associations, is something like a million-to-one. On the other hand, there is no young, compact cluster associated with SGR 0526-66. For this SGR, a physical association with the SNR N49 in the LMC still seems plausible.'

3. Burst and Transient Source Experiment.
4. Specifically, events unfolded thus: In their paper published in 1992 Duncan and Thompson made the connection of the theoretical magnetars with SGRs. Then in a paper published in 1993 (Thompson & Duncan 1993 Ap J 408, 194; sections 14.4 and 15.2) they suggested that the unusual X-ray star 1E 2259+586 was a magnetar. Finally, at a conference in March 1995 (with proceedings published in 1996) Duncan and Thompson suggested that the emerging class of X-ray stars resembling (and including) 1E 2259+586 which they called 'AXPs' were magnetars. This was eventually verified when the AXPs were found to emit SGR-like bursts. Now there are nine known AXPs, and a few more good candidates.
5. A common hand-held magnet has a field strength of perhaps a hundred gauss, while the strongest magnetic fields created in the laboratory are around 100,000 gauss.

14 'Seeing Double'

The initial observation took less than five minutes. A 23 millisecond pulsar 2000 light years away in the constellation of Puppis. Even from the first discovery observation, it was clear that the pulsar was in a short-period binary orbit. Follow up observations revealed signs that the companion was massive, perhaps even a neutron star. Binary pulsars had been discovered before – 80 had been found before this one, and there are now 134 known. The first resulted in Hulse and Taylor receiving the Nobel Prize (see Chapter 9). But while these systems involve a pulsar orbiting with a neutron star, the neutron star companions remained relatively anonymous: even if they are pulsars their beams do not intersect Earth and so remain invisible. Undetectable at the time of this pulsar's discovery was a much slower, second pulsar signal. If this was indeed a double pulsar, it would provide a unique test bed for the most stringent test ever of Einstein's theory of gravity, general relativity.

Discovery

The story begins with the discovery of what originally seemed to be yet another millisecond pulsar. The initial observation was made in April 2003 by Marta Burgay, a PhD student from Bologna. Burgay was one member of a team of astronomers from Australia, United Kingdom, Italy, India and the United States which was using the 64 meter Parkes telescope to survey the southern skies for pulsars.[1] Although the original observation took only a few minutes, in that time it showed a relatively large change in period due to Doppler shift: PSR J0737-3039 seemed to be part of a binary system and was added to the list of objects slated for follow-up observations. During a subsequent observation the Parkes telescope was trained on the pulsar for five hours straight during which time a detailed record of its behavior was recorded. The pulsar did indeed turn out to be a member of a binary, one with a 2.4 hour orbital period and a mildly eccentric orbit. The companion was also massive, at least 20% more massive than the Sun. With such a short orbital period, the diameter of the orbit must be correspondingly small; in fact it would fit inside the diameter of an ordinary star. These facts led the team to the conclusion that the companion was also a neutron star, and the system became one of only half a dozen double neutron stars known. Two questions remained, however. Could this neutron star be a pulsar, and if so was the second pulsar detectable? The answers to both these questions would be answered during a surreal episode of discovery that highlights once again the territorial nature of science.

An Inappropriate Discovery

Six months later, on October 18th 2003, astronomer Duncan Lorimer was at Parkes testing a new search technique he had designed to detect binary pulsars. He told me the following story. Lorimer was at Parkes on an observing run where typically several projects are scheduled over two weeks. The observers on duty therefore tend to often take data for other people, as well as their own projects. Lorimer was working hard on some code to 'barycenter' pulsar data, that is, resample the original time series so that the periodic signal present does not contain any unwanted Doppler shifts due to the Earth's rotation during the observation. The code works by using an ephemeris to calculate the arrival time at the Solar System barycenter which is, to a very good approximation, an inertial reference frame. This is important for long search observations of globular cluster pulsars, where astronomers typically integrate for several hours, and such Doppler shifts can become important.

However, the same ephemeris can also include orbital parameters of a binary system along the line of sight so that it is also possible to remove the Doppler shifts due to orbital motion during the integration. Now, PSR J0737A-3039 wasn't published at the time, but being based at Jodrell Bank, Lorimer was well aware of its existence, and had observed it on multiple occasions while at Parkes. 'It was one of the most compact and highly accelerated binary pulsars known, so it was a great test of my code,' he said. Lorimer set up the software to remove the orbital effects of the original pulsar given the known parameters. 'After a bit of debugging, I got it to completely remove the orbital effects so that the pulsar appeared as a signal of constant frequency. I went to bed that night happy that I'd achieved something with my efforts of the day,' he recalled.

Lorimer's next shift started around 4am. Overnight he'd had the thought that he could use his code to look for pulsar B, if indeed it existed. 'I wasn't supposed to be doing this at all, as I was not part of the project, but alone in the Parkes control room with the data and code. . .' Temptation – or more correctly curiosity – got the better of him. He looked at his search results for the data correcting for pulsar A's orbit and saw a marginally significant signal with a period of 2.8s. Still half asleep, he wasn't really sure what this was. 'My initial hunch was that it was just radio frequency interference; such signals are quite common. You get harmonics of airport radar and such with periods of a few seconds,' he said. Then he took the final step: Lorimer corrected the data for the orbital period of the companion, which amounted to just flipping the sign of the semi-major axis in the ephemeris! The signal that he'd originally seen in the data for pulsar A was now highly significant.

'At this point my heart was pounding,' he remembers. 'Once I folded the data at the correct period and looked at the profile it was obvious that this was another pulsar.' It was still only about 4:30am and he spent the next few hours refining the analysis in complete isolation from the rest of the world. Maura McLaughlin, Lorimer's wife and fellow astronomer, also present for this observing run, arrived at the telescope later that morning and the two set about measuring the orbital parameters for pulsar B and getting the first measurement of the mass ratio of this amazing system. The discovery of the first double pulsar was extraordinary for a number of reasons. Not only was this added confirmation of the fact that millisecond pulsars form in binary systems, here were two pulsars traveling around each other at a thousandth the speed of light. The system held the promise to test general relativity in ways never before possible. But science is terrestrial and territorial. It would soon be driven home to Lorimer that he had caught a rather large fish alright, but in someone else's pond.

What on Earth were You Thinking?

The same day, Matthew Bailes was on a 'rather smelly bus' headed for Parkes. Bailes had been Lorimer's Masters Supervisor. Bailes knew that Lorimer and McLaughlin were at Parkes, and called to see if they were able to pick him up from the bus stop. Not long into the conversation, Lorimer broke off the small talk by saying 'I've made a bit of a discovery,' and then proceeded to tell Bailes of the discovery of the second pulsar orbiting PSR J0737-3039. Bailes' first response was 'That's amazing. This is one of the greatest pulsar discoveries of all time! You're an absolute legend! And,' he continued, 'you're in all sorts of trouble! What on Earth were you thinking?' This was a major discovery, no doubt, but Lorimer had made the discovery using someone else's data. In effect, Lorimer had made a discovery of the first double pulsar without being a member of the discovery team. The following day, Lorimer and McLaughlin processed a large amount of archival data to establish the existence of the second pulsar. Once they were absolutely sure of the facts, Lorimer called Michael Kramer one of the team members at Jodrell Bank, waking him at 5am to tell him of the discovery. The political fall-out that ensued was tough. 'I had made a remarkable discovery on some data that I was only looking at for test purposes,' Lorimer said. His name did appear on the discovery paper, but it is quite conceivable that the double pulsar team could have left him off. 'When the big splash hit *Science* in January 2004 I felt simultaneously lucky to be involved, but also a bit hurt not to perhaps feature higher up the paper,' Lorimer told me. 'It certainly was a great experience, and when I look back on my career, the thrill of the discovery will be right up there on the reasons why I got into astronomy in the first place.'

There is no doubt that the barycentering code Lorimer developed helped him make the discovery. However, the pulsar was in fact strong enough to be seen using conventional techniques. 'It would have been seen earlier had it been looked for,' Lorimer commented. At the same time, the odds were stacked against Marta Burgay making the discovery herself: the second pulsar is only visible for about 20 minutes each orbit, and was invisible in the data she used to find the original millisecond pulsar. So let's now look at the nature of this extraordinary system.

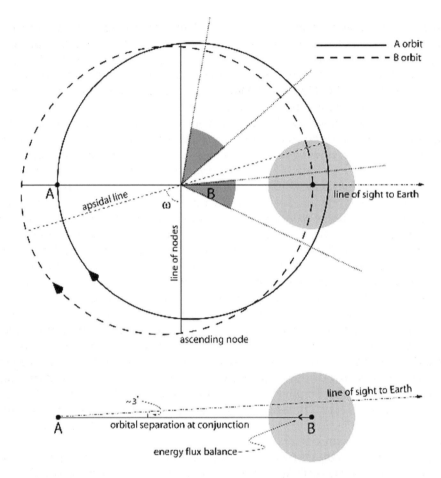

Figure 14 Configuration of the double pulsar PSR J0737-3039. Updated from an original picture which appeared in *Science* **2004. (Courtesy of Andrea Possenti.)**

How to Make a Double Pulsar

In Chapter 10 we saw how an ordinary pulsar can be accelerated up to millisecond periods by the accretion of material from a highly evolved companion. It seems this same process can lead not only to millisecond pulsars, but to binary pulsars. Let's recap. We begin with a pair of stars, one larger than the other. Having more mass, the larger star evolves more rapidly than the smaller companion. In time, the larger star will pass through its red giant stage and explode as a supernova. The result of this event is a pulsar. If the conditions are right, the pulsar will remain gravitationally bound to the smaller (but still massive) companion star. So far so good. As with the formation of other millisecond pulsars, the companion will eventually run out of nuclear fuel and evolve into a red giant. Matter from the swollen monster will spill over onto the pulsar, spinning it up to millisecond speeds. As it does so, it becomes an X-ray binary. The result is a millisecond pulsar sharing an orbit with a red giant. The similarities between previously discovered millisecond pulsars or even the binary pulsar and PSR J0737-3039A and B end here. What happened next was both dramatic and of inestimable scientific value.

The more massive star went supernova some 210 million years ago. Tens of millions of years later the smaller of the two also reached the end of its life and became a red giant, turning the original pulsar into a millisecond pulsar. While the smaller star was losing mass to the original pulsar, however, much of the matter was ejected from the system altogether carrying orbital energy with it. This caused the two stars to move closer towards each other. Stars start out as mostly hydrogen, and as they 'burn' their nuclear fuel the heavier elements – the 'ash' if you like – falls to the center of the star, resulting in an onion-like layering of elements of increasing weight closer to the center. As a result, highly evolved stars have an outer layer of hydrogen surrounding a layer of helium. But the outer hydrogen layers were shredded from the companion star by the gravitational interaction with the pulsar, leaving it with an outer layer of helium. As a result the star became what is known as a 'helium star'. Nonetheless, it was still a massive object and was destined to end its days as a supernova. Most of the time such a cataclysmic event would disrupt the orbit of the ancient binary and send the two stars on their separate ways. This is why so many millisecond pulsars are isolated in space. As the smoke cleared around this pair, however, it revealed something extremely unusual: PSR J0737-3039 had managed to cling to its smaller companion, which by now had become a pulsar in its own right. The former stellar binary had evolved into a double pulsar, one millisecond, one normal.

The two pulsars constitute an extraordinary system. They are separated by a tiny distance, just 900,000 kilometers, or a little more than twice the distance from Earth to the Moon. Traveling at 330 kilometers every second, the pair whip around each other every 2.4 hours. By great coincidence, not only are the beams from the pulsars both passing across Earth, the orientation of the orbits is such that it is edge on as seen from Earth. This means that once each orbit, one pulsar will pass in front of the other, obscuring its beam. Such an event is called an occultation, and means that astronomers can use the occulted beam to probe the atmosphere and magnetic field of the foreground pulsar. Further, the orientation of the rotation axes takes a mere 70 years to precess,[2] while the orientation of the orbit swivels around in a mere 21 years. This results in a continual shifting of the beams so that over time they will pass through different parts of the pulsars' magnetic fields and atmospheres. By studying the polarisation and dispersion of the beams, astronomers have been able to study the pulsar environment in detail.

For one thing, astronomers have been able to study the magnetosphere of a pulsar for the first time. Maura McLaughlin and collaborators used such observations to show that the magnetosphere of pulsar B was distorted by a stream of charged particles radiating outward from pulsar A. By monitoring the slow down rate of the pulsars, it was possible to determine the strength and hence size of the magnetospheres surrounding each pulsar. The magnetosphere surrounding pulsar A is a thousand kilometers across. Pulsar B, being a normal pulsar, is slowing down faster due to a stronger magnetic field which is a hundred times larger than pulsar A's. Since pulsar B routinely passes in front of pulsar A, it was predicted that the latter would be obscured for several minutes each orbit. When McLaughlin's team observed these occultations, however, the beams from pulsar A were only blocked for between 20 and 30 seconds. The explanation they came up with was that the magnetosphere of pulsar B is being distorted into a comet-shape by the intense wind from pulsar A. Whenever the two pulsars align, as happens each orbit, we see the smallest possible dimension of the magnetosphere, resulting in the brief occultation.

General Relativity on Trial

But by far the most exciting aspect of the double pulsar is its potential to test general relativity. As we saw in Chapter 9, the first binary pulsar allowed Hulse and Taylor to show that energy from the system is being carried away as gravitational waves, resulting in the two objects spiralling in towards each other at exactly the same rate predicted by general relativity. This is good news for gravitational wave scientists: clearly gravitational waves exist even if they haven't yet been directly observed. As intense as the binary pulsar discovered by Hulse and Taylor is, PSR J0737-3039 is an extreme test bed for general relativity. Astronomers quickly determined that just as the original binary pulsar is spiralling inward, so too is the double pulsar. As they emit gravitational waves, the two neutron stars are moving closer together at the rate of about 7mm per day, which means they will coalesce in approximately 85 million years, exactly as predicted by general relativity. This has an important implication for gravitational wave astronomy. Until now, the number of known double neutron stars was rather low, and of those known only a few are expected to merge in times less than the age of the Universe. Since merging neutron stars are an important source of gravitational waves, this left the chances of detection of gravitational waves rather uncertain. The discovery of the double pulsar improves those chances: because it is destined to merge in a cosmologically short time it implies others must be doing the same thing about ten times more often than previously thought.

But this is just the beginning of what PSR J0737-3039 has to offer. The concentration of the masses into neutron stars combined with their proximity to one another and their fortuitous orientation has created a bizarre region of space where spacetime is curved immensely, giving rise to measurable relativistic effects. Further, it is ten times closer than the binary pulsar discovered by Hulse and Taylor, making it much easier to observe.

All orbits are elliptical, with one long axis. Precession is caused by the intense warping of spacetime surrounding the two pulsars. When the astronomers analyzed the orbits of the double pulsar they found that it was precessing at some 17 degrees per year, four times faster than any previously known system. At this rate, the orbits will complete one rotation in just 21 years. Not only does the orientation of the orbit precess, so does the orientation of the rotation axes of the pulsars themselves. This is significant since, as the axes move, astronomers see different parts of the irregularly shaped beams. Changes in the shape of the beam show up as changes in the shapes of the radio pulses.

The elliptical nature of the pulsars' orbits gives rise to an even more important phenomenon involving changes in time. Because of the elliptical orbits, the pulsars periodically move closer to each other, subjecting them to different amounts of gravity. A useful analogy of spacetime is to think of a rubber sheet stretched flat. Whenever you place something with weight on the sheet it causes the rubber to deform around the object creating a depression with the object in the middle. Spacetime is analogous to this, and the more massive the object the steeper the sides of the depression. All objects in space, even you, deform spacetime to some extent, but more massive objects deform it proportionally more. Having so much mass concentrated into such a tiny space, pulsars really deform spacetime quite a lot, and the sides of the depression are steep indeed. But the key thing here is that it is not just space that is deformed, but space*time*. The closer you are to a gravitating source, the more time slows down. The situation with two orbiting pulsars is that you have two incredibly accurate clocks passing through spacetime that is very steeply curved. As they orbit one another, the pulsars will experience different changes to time – known as time dilation – depending on how close they are to each other, which varies as they move around their elliptical orbits. The highly precise pulses from the pulsars carry this information all the way to Earth where eager astronomers can study the changes in time close to the pulsars.

The most important test, however, comes from what is known as the Shapiro delay. This is the delay imparted to a signal as it passes through the curved spacetime of a massive object. As the pulses pass from one pulsar through the gravitational field of the other, they are delayed by the gravity of the second. The delay in the double pulsar is tiny, just 90 millionths of a second, but it is measurable. The results show that general relativity's predictions are accurate to within 0.05%. The double pulsar has provided the most precise test of general relativity to date.

The discovery of the double pulsar is significant to more than pulsar astronomers. Each December, the prestigious journal *Science* creates a list of the most significant scientific discoveries of the year. In 2004 the winner was the confirmation that planet Mars was once warm and wet, meaning it could have spawned life. Coming in at 6th place was the discovery of the double pulsar.

References

1. We will look at the details of this extraordinary survey in the next chapter.
2. The direction the axis of an orbit or rotation points swivels around in a specific and predictable fashion, a phenomenon known as precession.

15 'Of Multibeams and RRATs'

One of my first memories of pulsar astronomy was a remark made by Dick Manchester from the Australia Telescope National Facility. I asked him what it was about pulsars that could keep a scientist of his calibre interested in the same subject for decades. His response stuck in my mind: 'I think the thing that makes pulsar astronomy so interesting is that every few years something totally unexpected shows up.' As we've seen, the very discovery of pulsars could be described as serendipitous, but that's not a satisfactory description of what really takes place in pulsar science. Serendipitous is finding a dollar in the street while you're out walking the dog. What happened with the discovery of pulsars was the result of scientists like Antony Hewish creating an extra-ordinary new telescope, and the diligent and keen attentiveness of Jocelyn Bell to search through so much data looking for something that could easily have been passed over as interference. This is how it's been throughout the story of pulsar astronomy: curiosity matched with outstanding expertise and sheer hard work have resulted in an amazing array of discoveries of these dead and dying stars.

It is fitting, therefore, that the most recent discovery in the world of pulsars, the last discovery we will talk about in this book, followed similar lines. This discovery is not just of a new type of pulsar: it was the revelation of an entirely new breed of pulsar, one that may outnumber the conventional pulsars by four to one. These objects lack a glamorous name – they've been dubbed RRATs – and may even represent the final deaths of these stars that ended their first lives millions of years ago. Nonetheless, they are of such importance to our understanding of pulsars that even the discovery paper looks toward the emergence of new radio telescopes that will help astronomers find more of these objects. The discovery itself, however, was the result of an extraordinary instrument developed for, not pulsar astronomy, but an entirely different type of astronomical research. Called the Parkes Multibeam Facility, it was designed to survey the distribution of galaxies. But when Manchester and Andrew Lyne heard about it, they quickly realized its potential and utilized it in what has become known as the Parkes Multibeam Pulsar Survey. Ten years later their foresight has paid off: as well as revealing some of the amazing systems we have already explored in this book, it has doubled the number of known pulsars. It also led to the discovery of RRATs.

Parkes Multibeam Pulsar Survey

The main purpose of surveying the sky for pulsars is to try to establish how many are out there and what they're like. In addition, pulsars act as probes of the interstellar medium and the Galaxy's magnetic field. Many pulsars have been discovered at great distances in directions close to the center of the Galaxy. There have been many highly successful pulsar surveys at telescopes around the world, but none has discovered as many pulsars as has Parkes: in fact the Parkes radio telescope has found more than twice as many pulsars as all the world's surveys put together. 'I think that's pretty fantastic,' comments Manchester, adding that the Parkes contribution is 'a real mine of data'. As impressive as Parkes' 64 meter dish is, when compared with other telescopes it's nothing special. 'It's not a huge telescope,' Manchester points out. 'There are a lot of telescopes in the world much larger than Parkes.' So what's been the secret? Manchester points to three reasons. First of all is the telescope's location in the southern hemisphere: the center of the Galaxy goes overhead and this is where pulsars seem to be concentrated. Second, there has been a very good and experienced team of people doing the work. Many of these people we've met in the course of this story: Dick Manchester, Andrew Lyne, Matthew Bailes, Michael Kramer, and Simon Johnston.

The third major factor has been technology. All through this story of pulsars, improved technology has allowed the discovery of new objects and systems with unexpected characteristics. But there was one instrument destined for the Parkes telescope that was originally intended for an entirely different purpose: the Parkes Multibeam Receiver. The device has the size and outward appearance of a large water tank but went on to revolutionize pulsar survey work.

Multibeaming

A major limitation for all radio telescopes is their field of view. Whereas optical telescopes can observe 'large' areas of sky, doing the same thing at radio wavelengths isn't that simple. To build up an image of an extended source, an individual radio telescope has to be pointed at the sky, allowed to collect sufficient signal to form a point image, and then moved on to an adjacent point in the sky where the operation is repeated. Point by point an image is created like a pointillist painting. This tedious and time-consuming task is necessary because radio telescopes usually have only a single feed – the device which catches the incoming radio signals and turns them into electrical impulses – at the focus of the telescope's dish. As with optical telescopes, the larger the radio telescope's dish the fainter the object it can detect, but the smaller the patch of sky it can take in.

All that changed in January 1997 when the Multibeam Receiver was installed, giving the 64 meter dish a field of view about twice the size of the full Moon. Additional software and hardware was also installed at Parkes enabling astronomers to create real time, multi-wavelength images of the radio sky, a radio telescope's equivalent of a color photograph. The Parkes Multibeam Facility was the brainchild of Lister Stavely-Smith and developed by scientists and engineers of Commonwealth Scientific and Industrial Research Organisation (Australia's largest national research organisation), in collaboration with scientists from other Australian institutions, Jodrell Bank and the University of Cardiff. The Multibeam Facility was to have a total of 13 feeds allowing the Parkes telescope to observe 13 points on the sky simultaneously. The 13 feeds sit in the telescope's focus cabin perched high above the dish and collect the radio waves reflected by the dish below. The signals are then fed into an evacuated dewar mounted above the feeds which is cooled to 20 degrees Kelvin to reduce electronic noise. Inside the dewar are very low noise amplifiers which extract components of the radio signals at two linear polarisations, amplify them, and send them down into the control room.

Once in the control room, the signals are channelled through a bank of electronics called the correlator. This device produces 13 individual radio spectra, each corresponding to one of the 13 beams, of any object falling within the telescope's field of view. The correlator allows the astronomers to observe celestial radio sources at several wavelengths simultaneously, a capability which has important applications in mapping the nearby Universe by measuring the redshift and hence distance to galaxies. Measuring the redshift of a galaxy to determine its velocity and hence distance depends on identifying known spectral lines and calculating their (redshifted) wavelength. In optical spectra the entire visible spectrum is laid out for relatively easy identification of spectral lines. Things aren't as simple at radio wavelengths, however, because radio telescopes generally observe one wavelength at a time. Since the wavelength of a spectral line depends on the velocity of the galaxy, astronomers have to go searching for the line before they can calculate the galaxy's redshift. What the correlator allows is relatively fast analysis of a wide band of the radio spectrum in real time so that specific lines can be identified, allowing astronomers to determine the velocity and distance of a large number of galaxies with much greater efficiency. One of the Multibeam Facility's first tasks, in fact, will be to map the sky at the 21cm line of neutral hydrogen, the so-called H I line. The H I line is important because it allows astronomers to measure the red shift, and hence velocity and distance, of galaxies beyond the Milky Way.

Surveys of distant galaxies have been already been carried out at optical wavelengths, but there are regions in the sky where galaxies are invisible at optical or even infrared wavelengths. 'A region of special interest in the southern sky is the galactic plane,' explained Staveley-Smith. 'While we can look above and below the plane of the Milky Way with ease, looking through the central bulge of the Galaxy is possible only at radio wavelengths, and there's some very interesting structures that are being hidden by the Milky Way.' One of the main uses of the Multibeam Facility was to explore this previously hidden region of sky, historically known as the 'zone of avoidance'. Astronomers made use of the fact that the 21cm line pierces straight through the Milky Way to make a much more detailed map of the region. Astronomers hope that this survey work will pick up numbers of previously unseen galaxies rich in neutral hydrogen, such as the so-called 'crouching giants' – large galaxies with very low surface brightness that have escaped detection so far.

Piggyback Astronomy

While the Parkes Multibeam was designed for HI surveys, two astronomers realized it would be ideal for pulsar surveys. 'Andrew Lyne and I immediately realized this would be a good instrument for a pulsar search,' recalls Manchester. The H I frequency is ideal for searching for pulsars along galactic plane and such a search could be piggy-backed onto the proposed Multibeam H I survey. The effects of interstellar dispersion which is high in the plane is diminished by the higher frequency of the survey: the higher you go in frequency the less dispersion and scattering by interstellar gas and dust matter. At the same time, pulsar signals become weaker at higher frequencies; pulsars have steep spectra means they are weaker at higher frequencies. '1400 MHz is a good compromise,' Manchester points out. 'The scattering and scintillation are not too bad and yet the pulsars are still relatively strong.' Manchester and Lyne saw the potential of the Parkes Multibeam in the early development stages of the instrument, but also realized specifications for HI survey were not ideal for pulsar surveys. When searching for pulsars, sensitivity is essential: the number of pulsar discoveries is proportional to the sensitivity – double the sensitivity and double the number of pulsars. And sensitivity is related to bandwidth. The original design called for a bandwidth of 150 MHz, but this was too narrow for pulsar surveys. Manchester and Lyne convinced the Multibeam team to increase the bandwidth to 300 MHz (of which they ended up using 288 MHz). Getting that extra bandwidth meant they would find 1½ times as many pulsars. 'In fact it's a little better than that. Sometimes pulsars are strong at one end of the band and not the other, and so you just end up with more chance of picking up these pulsars.'

International collaboration – the pooling of expertise and resources of people from around the world – is essential to the success of modern surveys. The amplifiers were redesigned to provide for a greater bandwidth. Lyne then organized for people at Jodrell Bank with help from engineers at Bologna to design an analog filter bank that had 13×2 (for two polarisations) separate filter banks. Each filter bank had 96 3-Mhz channels. 'It's a big machine,' Manchester observes, taking up three racks at Parkes. Meanwhile the ATNF built the front end and Fernando Camilo, then at Jodrell Bank, came to ATNF for six months to write the front end data acquisition program. Manchester set about writing the control program for the system with help from Lyne. 'So much effort is required these days in putting such an experiment together, operating the telescope, processing the results and doing the science, that you absolutely need a big team of friends to do the work and several institutions to share the resource provision.' The whole project came together in 1997 to take immediate advantage of the recently installed Parkes Multibeam Receiver. The system was commissioned over a few weeks and the survey began in mid-1997.

Wishes Come True

The Parkes Multibeam Pulsar Survey began in August 1997, and with a sensitivity seven times greater than anything previously attempted there were high expectations of a very successful search. They weren't let down. It was anticipated that somewhere between 300 and 400 pulsars would be discovered; the survey soon turned up over 750 pulsars, including a further 15 millisecond pulsars. By November 1998 the 1000th pulsar known was on the board. Data processing was done on a variety of computers at ATNF, Jodrell Bank, Bologna and MIT. 'Every night we'd try and get all the workstations in the building – some fifteen to twenty computers – to work on this project while they weren't working on anything else,' Manchester recalled. New pulsars are being found even up to the time of writing by using improved algorithms to analyze the original data. The Parkes Multibeam Receiver is now responsible for discovering more than half the known pulsars. The total number of pulsars found with Parkes is now two thirds of the known pulsars.

Such a record hasn't gone unnoticed, nor unchallenged. For example, when astronomers at the largest radio telescope in the world, Arecibo, heard of the success of the Parkes Multibeam Pulsar Survey, their first response was 'How the hell did they get so far ahead of us?' Their next response was 'Well we'd better get them to make one for us.' And so the ATNF made a 7 beam system for Arecibo which is to be used for the P-ALFA survey. 'We definitely had the jump on the rest of the world and made good use of it,' said Manchester.

Still More Discoveries

Among the myriad discoveries made with the Parkes Multibeam – which includes the double pulsar PSR J0737-3039, the 'Jewel in the Crown' of pulsar science – Manchester and his colleagues have also found a new population of slow, young pulsars with incredibly strong magnetic fields. They seem to be related in some way to magnetars in that they have similar rotation properties: a very rapid spin down rate (which reveals their strong magnetic fields) but rather long periods. But what is surprising about these objects is that their magnetic fields as strong as those of magnetars, and yet they are totally different in that they emit at radio wavelengths and not at X-ray wavelengths. (Most magnetars, you'll recall, emit only in X-rays; only in the last year or so have magnetars been detected at radio wavelengths.) These pulsars, however, are ordinary pulsars that shine at radio wavelengths.

We are nearly at the end of the story of pulsars as it can be traced to date. There is no question that discoveries will continue to be made: some with new

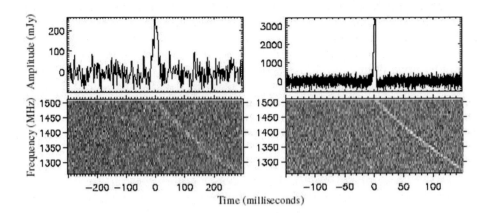

Figure 15 RRAT bursts from J1443-60 (left) and J1819-1458 (right). The lower panel shows the dispersed nature of the bursts detected in the individual frequency channels. The upper panel shows the de-dispersed signals, obtained by summing outputs of the individual receiver channels at the optimum value of the dispersion measure. McLaughlin *et al.* 2006, *Nature*, 439, 817. (Courtesy Maura McLaughlin.)

observations yet to be made with new instruments; others through the re-analysis of old data using new software. In the last chapter we will look to the future of pulsar astronomy, which is likely to be revolutionized with the use of a yet-to-be-built radio telescope of unprecedented dimensions. But there are still major discoveries to be made using contemporary instruments, and the most recent is the discovery of RRATs.

Oh, RRATs!

There is a type of radio signal that has not yet been well-surveyed: transient signals that last only a brief time and so may be missed during conventional radio sky surveys. During the PMPS, the telescope was aimed at one patch of sky after another for 35 minutes each. In late 2001, Jodrell Bank astronomer Maura McLaughlin arrived at Jodrell Bank Observatory to work with Lyne's group and suggested adding a new search algorithm to sift through the data for isolated dispersed pulses. Over the next few years, she began reanalysing data from the survey collected between January 1998 and February 2002. Each of the observations had been carried out near the plane of the Galaxy. She was searching for transient signals when she, Andrew Lyne at Jodrell Bank, and other colleagues found a total of 11 sources that had been initially overlooked. Once again, these transient signals might have been artificial interference were it not for the fact that the dispersion measure of the bursts from any given source was the same, placing each source far from the Solar System in the galactic plane. What was strange, however, was that when follow-up observations were made, the signals were found not to be one-off events, but recurring signals. Each burst lasted anywhere from 2 to 30 milliseconds, with pauses that last from minutes to hours. In fact in a given day the total amount of radio emission might last for only a second, with anywhere from a few to a few hundred radio pulses observed from the various sources. Additionally, they were radio bright, in fact brighter than the individual pulses of most pulsars. 'Because of their repeatable nature (i.e. multiple bursts detected at the same value of DM), we first named these sources Repeating Radio Transients (RRATs),' McLaughlin explains. 'We soon realized, however, that we could determine underlying periodicities for some of the RRATs by calculating the greatest common denominator of the differences between the bursts.' From the pulses received, McLaughlin and her colleagues were able to identify periodicity in the signals ranging from 0.7 to 7 seconds. From this periodicity they inferred that the objects must be rotating, giving rise to the final class of pulsar: rotating radio transients, or RRATs.

RRATs are difficult to find; after all, they're silent most of the time. This makes them ephemeral objects that, given any observing session, are unlikely to show up. But this in itself has an extraordinary implication: if they show up in such numbers during such a brief observation, and they don't reveal themselves all that often, then there must be an awful lot of them. Take the source J1911+00, for example. This object bursts every three hours, which means that there was a 20% chance of finding it in the 35 minute observing window during the PMPS. Extrapolating from the area searched so far, and there is a large number of RRATs out there waiting discovery. In the Galaxy, there's an estimated 100,000 pulsars; based on the discoveries so far, the number of RRATs may be as high as 400,000. The discovery of RRATs has begun a new chapter in the ongoing story of pulsars.

So What Are They?

There have been two ideas put forward to explain the nature of RRATs that we'll look at here, although it is undoubted that, even as this book is being printed, new theories will be put forward. Let's look at the facts. The periods demonstrated by RRATs are long, anywhere from 0.7 to 7 seconds. This is typical of pulsars on their last legs. After millions of years, they have slowed down to the point where their emissions are weak and their rotation periods long. They are all lone objects: none appears to be associated with a stellar companion. They are silent most of the time, only occasionally and briefly sending out a beam of radio waves.

In a 2007 paper published by Bing Zhang, Janusz Gil and Jaroslaw Dyks, two models for RRATs were discussed. One is that RRATs are normal pulsars near death. Although the origin of pulsar emissions is not clear, it is believed that it relies on electron-positron pair production near the surface of the neutron star. When this pair production ceases, so do the radio emissions, and the pulsar is essentially dead. But this death is not instantaneous: pulsars do not suddenly switch off. Rather they slow down gradually, and so it follows that they must enter a stage when the radio pulses begin to falter. More and more the radio emissions fade, until the star is undetectable. However, the whole radio emission mechanism is linked with magnetism, with the beams emerging from the offset magnetic axis of the neutron star. Now anyone who's ever seen the Sun with a telescope will likely have seen sunspots. These dark blotches are in fact magnetic storms on the surface of the Sun. If a sunspot-like magnetic field on the surface of a neutron star drifted near the star's magnetic pole, the neutron star might briefly emit radio beams once more. This could only happen if the pulsar were near death, not well beyond it. As Zhang, Gil and Dyks put it, 'many pulsars not deep below the death line could jump out from the graveyard occasionally...'.

There is an alternative to this scenario, however. In the same paper, Zhang, Gil and Dyks point out that there is a type of pulsar which very occasionally and unpredictably misses a beat. One example is the pulsar PSR B1822-09. Normally, this pulsar emits regular pulses, but occasionally it drops out. One theory suggests that such a 'nulling pulsar' may be brought about by a reversal of the direction of the radio beam. Every now and then the beam, instead of aiming outward across Earth, reverses direction and passes near the pulsar itself in the opposite direction. Since the beam has to be passing across Earth in order to be seen, the geometry of the reversed beam prevents us from seeing it. Hence, from time to time, the pulsar flash is missing. It follows that, seen from the opposite direction, occasionally a beam will appear to be coming from an otherwise quiet neutron star. The main beam is normally invisible, and we only see the pulsar when this beam reverses direction and is aimed in our direction. An alternative theory suggests that the RRAT emission is due to a belt of debris surrounding the pulsar.

McLaughlin points out that no two RRATs are the same. Some have a clear on and off phase while others have random isolated bursts. Some are very extreme nullers, while others are very similar to normal radio pulsars. While several explanations have now been put forward to explain RRATs, it is likely that different explanations are needed to explain the wide variety so far observed. Further, McLaughlin points out that the discovery of RRATs and the difficulty encountered in explaining them 'highlights our still very poor understanding of the radio pulsar emission mechanism'.

How Many RRATs?

Just how many are out there has been estimated by looking at how many were found. For example, of the RRATs found so far, on average they emit a single burst every three hours. The first RRAT discovered was found in a 35 minute observation, meaning there was a 20% chance of discovering it. Add to this the difficulty of finding RRAT signals in all the noise, the fact that the first eleven RRATs were found by visual inspection of chart recordings by a single observer, and that the sensitivity of searches so far did not extend to looking for pulses of greater duration than 32 milliseconds, and there are reasons to suspect there are a lot of RRATs out there awaiting discovery. When McLaughlin and her colleagues included these and other factors, they calculated there could be as many as 200,000 RRATs in the Galaxy. This is a large number of undiscovered pulsars, especially considering that the number of active 'normal' pulsars is estimated to be around 10,000. As much as astronomers thought they knew about pulsars, it may turn out they have just been looking at the tip of the iceberg.

McLaughlin and her colleagues are now in the process of re-examining observations made during previous surveys looking for these transient sources. When the Parkes observations in which the original 11 RRATs were discovered was re-examined, a further six RRATs were found. Similar searches of other survey records are revealing more and more RRATs. The discovery highlights the need for keeping an eye out for transient signals, and that's just what astronomers want to do. In fact the next generation of radio telescope promises much for the discovery of RRATs and other pulsars. It is to this latest chapter in the story of pulsars that we now turn.

A NOTE ABOUT PULSAR SURVEYS

There is simply not enough room in a book like this to mention all of the surveys and astronomers who have contributed to this field. However, while this chapter focuses on the Parkes Multibeam Pulsar Survey, it should be noted that there have been other surveys conducted at Parkes and elsewhere. Important surveys include the Swinburne survey by Bailes and his collaborators which covered a flanking region of the galactic plane. The double pulsar was in fact discovered during a high latitude survey as it crossed the plane and was not from the main survey. There have been many surveys of globular clusters and sources detected with the Energetic Gamma Ray Telescope on board the Compton Gamma Ray Observatory which have turned up many new millisecond and binary pulsars. Finally there was a survey of the Magellanic clouds which Dick Manchester and his colleagues conducted. While searching these data for RRAT-like objects, Lorimer and his colleagues found a new type of radio burst which they believe to be extragalactic.

16 'The Future'

The thing that makes you wonder is how astronomers and theorists will cope with the deluge of discoveries that are likely to come from the next generation of radio telescopes. Up until now, pulsar astronomy has thrived on a combination of ingenious technology and equally ingenious exploration and interpretation of data. No example is clearer than the achievements of the Parkes radio telescope that we looked at in the last chapter: through technology, ingenuity, experience and expertise, this telescope has made a greater contribution to pulsar astronomy than any other single observatory. But no matter how many modifications are made to existing telescopes, eventually their limitations begin to show. This is merely a limitation imposed by the collecting area of the radio telescope. In astronomy, bigger is better, and there is now a pressing need for a bigger telescope. But not just one with greater collecting area, this will be a new type of telescope, one that has unprecedented sensitivity and is backed up by technology not yet developed. The answer to this call has a name: the Square Kilometer Array.

SKA

SKA will be the largest radio telescope in history, dwarfing anything now in existence. Press releases breathlessly foretell its construction and its potential, and with justification. After all, when it's finished, SKA will unleash astronomers on a universe far beyond what they can presently reach. With a collecting area of a square kilometer – that's a million square meters[1] – SKA is expected to revolutionize our understanding of the distant, the intense, the old and the small. It will reach in every conceivable direction of time and space, exploring every astrophysical and cosmological phenomenon the Universe has to offer. Included in this new realm are pulsars. SKA not only has an unprecedented potential to discover more pulsars – both normal, millisecond, and RRATs – than ever before, it also promises to reveal pulsars whose existence is speculative and widely anticipated: the elusive pulsar orbiting with a black hole. Such a discovery will lead astronomers to an ever deeper understanding of the Universe. SKA will find the pulsars, and pulsars will reveal the nature of space and time.

What a Telescope

SKA will be a telescope of extraordinary proportions. Despite the name, it is a mistake to think SKA will be a telescope of a mere 1 km to a side. The 'square kilometer' in the name refers to the total collecting area of the individual antennas. The antennas, in turn, will be distributed across distances that span **a continent**. The reason for this is simple: the greater the distance between the dishes, the greater the resolution of celestial objects. Constructed as an interferometric array, SKA will consist of widely spaced antennas that can simulate the resolution of an immensely larger telescope. The antennas will be grouped into stations – up to a few hundred – each with a diameter of between 100 and 200 meters. Half of the collecting area will be concentrated within a region five kilometers across. Another quarter of the antennas will be within a radius of 75 kilometers, while the remaining antennas will range up to 3,000 kilometers from the inner array. Interferometric arrays aren't new: existing telescopes such as the Australia Telescope National Facility or the Very Large Array have been making extraordinary advances for decades, while the principle of linking isolated radio telescopes to simulate much larger instruments has been implemented not only on continental but also global scales and beyond. What SKA will do, however, is combine the principles of interferometers with a massive collecting area and technologies not yet developed.

So just how do you construct a telescope with a million square meters of collecting area? There are a few ways, but they fall into two main camps: a collection of individual dishes, and the use of flat phased arrays. In the first design, the telescope's antennas would be parabolic solid or wire mesh dishes that can be aimed at any part of the sky. With a diameter of 12 to 15 meters, depending on which design is chosen, a total of 4,400 dishes would be needed to fulfill the sensitivity specification of the telescope. Each dish could be linked collectively, or groups could be used as arrays to look at a number of objects simultaneously. An alternative to parabolic dishes is the use of flat phased arrays. This innovative approach has been pioneered by Dutch engineers through the Thousand Element Array (THEA) project. Each of the THEA tiles is a square meter consisting of 64 wide band antenna elements; SKA would call for over 50 million receiving elements. Steering of a telescope constructed in this fashion would be done electronically.

An International Effort

No one country can undertake such an immense project, and so SKA is made up of a consortium representing 17 countries, including Australia, Canada, China, Germany, India, Italy, the Netherlands, Poland, Sweden, the United Kingdom, and the United States. Of course, then the question arises: where to put it, and how do you decide? After access to the most important parts of the sky, the most important considerations include radio quietness (free from interference from radio, television, mobile phones, etc), access to technical and scientific resources, and the physical nature of the site. At the time of writing (June 2008) the location for SKA had been narrowed down to two sites: Western Australia and South Africa. A decision on which site will be the final home of SKA is due some time in 2009.

What to Look at?

SKA will concentrate on five key projects.

1. Cradle of Life: looking for life elsewhere in the Galaxy. This involves looking for Earth-like planets orbiting other stars, perhaps even detecting signs of intelligent life.
2. Probing the Dark Ages: studying the first black holes and stars that emerged after the Big Bang, with the aim of understanding what happened during those first dark times when the Universe existed but was not yet illuminated by stars.
3. Origin and Evolution of Cosmic Magnetism: how does magnetism influence the evolution of galaxies and stars, and what keeps these magnetic fields going.
4. Cosmology, Dark Matter and the Evolution of Galaxies: everything from the nature of dark matter and dark energy to the origin and evolution of galaxies.

But the most exciting project for the present story is this one:

5. Strong Field Tests of Gravity using Pulsars and Black Holes.

As we've seen in the preceding chapters, pulsars have offered very stringent tests of general relativity and what it has to say about gravity. So far Einstein's theory has passed every test thrown at it. But is it really the last word on gravity? One problem with testing general relativity using instruments either on Earth or even within the Solar System is that the gravitational field is relatively weak. The truth is that the Earth is surrounded by a relatively feeble gravitational field, and this places limits on what kinds of gravitational experiments can be carried out. Strong field tests are only possible with pulsars, and as we have seen they have provided some of the most stringent tests so far. The discovery of between 10 and 20 thousand pulsars – including up to perhaps a thousand millisecond pulsars – scattered throughout the Galaxy will constitute an immense and highly sensitive array of super-accurate time keepers with SKA at the center. This would form an ideal 'instrument' for studying gravity, in particular detecting gravitational waves with frequencies in the nano-Hertz range. By combining these observations with those made by Earth-based gravitational wave detectors such as Advanced LIGO (Laser Interferometer Gravitational wave Observatory) and LISA (Laser Interferometer Space Antenna). High-accuracy (<100 nanoseconds) timings of so many pulsars, SKA will help astronomers answer questions such as: Is Einstein's theory the last word on gravity? Are its predictions for black holes correct? Is the Universe awash with gravitational waves? Such an array of pulsars could even reveal the presence of high frequency gravitational waves left over from the Big Bang itself.

Within the thousands of pulsars that SKA is expected to find lie virtually every possible outcome of the evolution of massive stars and binary systems, including more than a hundred compact relativistic binary systems, and several more double-pulsar systems. Astronomers also expect to find the long sought after pulsar-stellar black hole system. General relativity makes quite specific predictions about the nature of black holes. Such a system would provide the most discriminating test of relativistic gravity so far, or for that matter foreseeable. We have already seen that the spectacularly successful surveys conducted using contemporary radio telescopes have unveiled some truly strange objects. One can only imagine what strange beasts lurk in the jungle to be explored by SKA. But the really exciting possibility is the possibility of finding a pulsar orbiting the supermassive black hole thought to exist in the center of the Galaxy. By observing such pulsars various characteristics of the black hole – mass, spin and quadrupole moment – can be determined and compared with Einstein's predictions.

Construction of the SKA will hopefully commence some time in 2011. By 2020 the telescope should be in operation, giving us a glimpse of a universe we have not yet imagined. That's ten years away, and ten years is a long time in pulsar astronomy. As we have seen repeatedly, the most extraordinary discoveries have been completely unexpected. With that in mind, it is impossible to know what the next chapter of this book should be. It may well be that before it reaches the bookshops, a new development, a discovery, a revelation will emerge that is too late to be included. We can only watch and wait.

Conclusion

We have now spanned more than a century of science: From the discovery of the neutron, the prediction of the neutron star and the spectacular discovery of pulsars, a new and totally unexpected realm of astronomy arose to challenge our understanding of the nature of matter, gravity, time and space. The emergence of pulsar astronomy was impossible to predict, as was the importance it has played in our understanding of the Universe. Despite this, if you ask an average member of the public what comes to mind when thinking of astronomy it's unlikely you'll hear the word pulsar. One possible reason why pulsars are relatively unknown among the public may be the lack of visual impact pulsars have: tiny, white and, at least at first glance, dull, pulsars do not have the public appeal of glossy photographs of planets, nebulae and galaxies. Most people can appreciate the Universe only in qualitative terms, usually visual. At least you can see a picture of a planet.

I believe the problem goes deeper, however. It is possible to explain to a non-scientist what a pulsar is, and the professionals working in public outreach do an outstanding job of doing so. However, once the non-scientist grasps the basics of what a pulsar is – a spinning neutron star – there is very little appeal. For most people, beyond the graphics showing a turning sphere emitting beams like a lighthouse, little is comprehensible: beyond lies a vast dark chasm of mathematics and physics that few will ever be able to comprehend. Some brave souls delve into general relativity and quantum mechanics privately, gaining a deeper than normal insight into the implications of the high density matter that is pulsars. But most people, rather than spoiling the view, step back from the mathematical precipice thus avoiding the complex notions and ideas that are at the very heart of pulsar astronomy. For most people, the mathematics and physics needed to truly appreciate the beauty of pulsars are far beyond their reach.

There is some hope, nonetheless. I am not a professional scientist, nor a mathematician; the physics and mathematics that pulsar astronomers work with is regrettably beyond me. But with the help of astronomers from around the world I have been shown that it is possible to appreciate the beauty, the stunning implications, the astounding behavior of nature that is revealed through the detailed study of the tiny spinning orbs that are pulsars. Rather than a chasm, I now see pulsar astronomy as a mountainous frontier beyond which lie horizons of such beauty they give us reason to go on. Astronomers, to me, are like mountaineers that have scaled the heights of mathematics and physics and seen what lies beyond. Sometimes they can show us photos; sometimes only descriptions. But they can see, and they call back down to tell us what they have learned.

Through the story of pulsars we have also seen that climbing those heights is not a serene activity: it is sheer hard work. Some falter, others find a hidden but easier path and streak ahead, often there is conflict. But what these people see is more important, I believe, than anything else humans undertake. Astronomy, I was once told by an astronomer I interviewed in the dome of the Anglo-Australian telescope, is a cultural activity. It offers no practical benefits; it doesn't feed hungry people or help us cure disease. Admittedly there have been important benefits in terms of imaging and other technologies that have come out of astronomy, but the bottom line is that it exists to satisfy human curiosity. Astronomy is a cultural activity, like art or music. But there is a huge and crucial difference that sets it far higher than any other human activity: it progresses. There is no progress in other human endeavours, only reshuffling of old ideas, colors, shapes and patterns. Scientists on the other hand advance our knowledge and understanding of the very nature of the Universe, and hence of ourselves. These are the true heroes of civilization. I hope that at least some of the excitement of pulsars, of the beauty of nature, of the stunning implications for the nature of matter and the Universe has been conveyed in this story.

Reference

1. For comparison, Parkes has a collecting area of 3,217 square meters, while Arecibo has a collecting area of around 40,000 square meters. SKA will have a collecting area a hundred times that of the Very Large Array.

Appendix: Recollections of a Scientific Error

Roderick Willstrop

Introduction

The oldest telescope at the Institute of Astronomy in Cambridge is the Northumberland Telescope, built between 1836 and 1839, and now used for public observing and by members of the University Astronomical Society. This telescope is famous, or perhaps infamous is a more appropriate word, for having been the telescope with which James Challis searched for the planet predicted by John Couch Adams, who saw it on up to three occasions, but failed to recognize it.

One might have expected that the continued presence of this telescope would serve as an 'Awful Warning' to all members of the Department to avoid making a similar error ever again. Unfortunately, it did not; Dr W. H. Steavenson, the last amateur astronomer to have been President of the Royal Astronomical Society, had installed his own 30-inch reflector in the grounds of the Cambridge Observatories in 1939, and from 1945 he continued a programme of visual observations of the brightness of a number of remnants of novae (including DQ Herculis) which he had started many years previously with a smaller telescope (Steavenson 1938, 1939, 1946, 1947, 1948, 1950, 1953). Unfortunately he looked at these objects only once on any night, and so missed the regular fluctuation with a period of 4 hours 39 mins found by M. F. Walker (1954). The following story illustrates how it came about that the Institute of Astronomy now has a 36-inch telescope with which the Crab Nebula pulsar was not discovered.

Background

The discovery of pulsating radio sources with periodicities of the order of 1 second was published in *Nature* on 24 February 1968. This caused enormous interest, indeed excitement, at the Cambridge Observatories, not least because the discovery had been made by our colleagues in Radio Astronomy, based in the Cavendish Laboratory. Hewish *et al.* (1968) put forward 'A tentative explanation of these unusual sources in terms of the stable oscillations of white dwarf or neutron stars.'

Cambridge Observatories

Attempts to make an optical identification of one of the pulsars were begun at the Cambridge Observatories within six weeks of their publication. John Jelley of A.E.R.E. Harwell brought equipment which included a Laben analyzer which we used in multiscaler mode to record the output of a simple photometer mounted at the prime focus of the 36-inch reflecting telescope. Observations were made in the early hours of 5th and 6th April 1968, but no evidence of optical flashes from CP 1919 was found (Jelley & Willstrop 1968).

The multiscaler used in April 1968 could only be set to search for one pulse frequency at a time. With the discovery of other objects having different pulse rates, a more versatile method of recording and an improved 'Mark 2' photometer were both desirable. I designed a photometer in which a wheel with four filters, approximately U, B, V and 'white', could either be rotated continuously to measure U-B and B-V colours more reliably than when the filters were changed at longer intervals, or the wheel could be stationary. My old notebooks have fewer dates recorded in them than I should now like, but it is clear that construction of this photometer was in progress in the Workshop in late July 1968 when I drew a circuit diagram for the controls of its filter wheel.

E. J. Kibblewhite designed and built a photometric recording system which used a high-speed punch for output. This could perforate 7 track paper tape at up to 110 rows per second. The actual rate of punching the tape was controlled by a 1MHz quartz crystal, whose pulses were divided first by 10, then by a series of binary scalers, adjustable so as to produce rates from 97.66 down to 1.526 punch operations per second (10 to 16 binary steps).

Pulses from a photomultiplier used in photon-counting mode were recorded in binary notation on the paper tape. More binary scalers were used to adjust the numbers sent to the punch; if these were greater than 63 the most significant bit would be lost, so the scaler was normally adjusted to produce output in the range 25 to 50. The tape was allowed to fall, or perhaps the word 'spew' was more appropriate when the punch was run near to its maximum rate, into a wooden box (a tea chest, about 60 cm cube) and at the end of each observation the tape was wound up using an adaptation of a hand-powered grinding wheel. The paper tape could later be taken to the Cambridge University Computer Laboratory and analyzed for any periodicities.

A conference was held in Cambridge on the instrumentation of the Anglo-Australian Telescope on 11th and 12th August 1968. Some time after this conference, but before 5th September, there is a note that the photometer was not operating correctly at 97.66 readings per second: the seventh track on the paper tape was used for a parity check, but at 97 readings per second about ten per cent of the rows on the tape had the wrong parity, and about 2 per cent of the readings were repeated. Some faults in the wiring of the photometer were found and it was hoped that these punch errors had been corrected.

While this equipment was still undergoing proving trials, an IAU Circular arrived, announcing the discovery by Staelin & Reifenstein (1968) of two pulsars in Taurus. As IAU Circulars are now sent around the world in a few milliseconds by email, it is difficult to remember that back in 1968 they were printed on small cards, and sent by airmail, so that they were received 3, 4 or 5 days after the nominal date of publication. This Circular was published on 6th November 1968 and would have been received on Monday, 11th November. The position of NP 0532, the nearer of the two pulsars to the Crab Nebula, was initially uncertain to within 3 minutes of time in R.A., and two degrees in Declination. Its period was given as less than 0.13 seconds, and the pulse width less than 120 ms.

Time on the 36-inch telescope had already been allocated to me, as I was planning to use the new photometer to look for pulsations in the light of white dwarf stars. This run started on 16th November and several nights were clear enough to allow some observing. An important feature of searching for regular pulsations in the light of stars or other objects is that fluctuations in the transparency of the sky are most unlikely to be regular enough to cause confusion. On 23rd November 1968 I had gone to bed at 21.30, but the sky cleared by 22.30, and I went out to observe.

The notebook records:

> *At 23.36 decided to start on Crab Neb. Found it with no difficulty. Can see faint starlike condensations. If I can find a photo I might identify the central object. Went to library at 23.55. Cloudy at 00.10 !! Carried out various tests. Reopened at 00.36 Clear patch. M1. 1 mm diaphragm, no guiding. Start at 00.45. Stop after 50 seconds. We are looking through cloud. Raining at 01.02. Cleared at 02.15. Reopened, set on AGK2 +21 541, then offset 2 minutes to Crab. Noted the star field, and just identified these stars on Sky and Telescope vol. 28, p. 334, 1964 by 02.30. 02.36 Saturn? Visible in W. hope!*
>
> *M1 02.56.30 > 11.5 mins*
> *96/second Prescaler 4 Typical reading 15 or 16*
> *1 mm diaphragm – no guiding.*

Later the same night, and again in the early hours of 25th November, further observations of the Crab Nebula were made, using 48 rows of tape per second, because Ed Kibblewhite and I were still not completely confident that the equipment was working exactly as it was intended to do when it was run at the highest rate of 97.66 readings per second. The number of photons recorded in the most significant 7 bits of the counter was transferred to a buffer store at intervals of 10.24 msec, and retained there for up to 9.1 msec while waiting for the punch to be ready for the next punch cycle. When the punch sent a signal that it was ready, the number in the first buffer store was transferred to a second buffer store, and the first buffer was cleared in readiness for the next set of data bits. Further tests after the Crab observations showed that our fears were groundless; the faults that had been present in August had been rectified and the data were being recorded correctly.

During that week there were five nights that were clear enough at some time or another to make observations of white dwarf or other stars, such as DQ Her, Wolf 1346, Hertzsprung 3, and Oxford +25 6725, and I spent my time in the following weeks feeding the paper tapes to the Computer Laboratory where I used the Titan computer with an inefficient program of my own to search for periodicities in the data. At that time I was not aware of the Cooley-Tukey Fast Fourier transform algorithm.

Meanwhile, other radio observations of NP 0532 were reported. Lovelace, Sutton & Craft (1968) in an IAU Circular published on 18th November and received in Cambridge on 21st November, gave the correct period of 33.09114 msec, and a substantial improvement in its position, which was given as 'within 10 arc min from the centre of the Crab Nebula'. The photometer had a field of view 4 arc min (5 mm) in diameter for finding, and diaphragms with diameters of 1, 0.5 and 0.25 mm (50, 25 and 12.5 arcsec). The bright sky in Cambridge made it impractical to use the 5 mm aperture for photometry of faint objects, and the area of sky to be searched was so much larger than the angle covered by the 1 mm photometry aperture that it seemed to be a very long shot to expect to be able to find anything.

On 1st January 1969 I compiled a list of questions that I should ask at Herstmonceux before applying for time on the Isaac Newton Telescope, and on 13th January I put my photometer on the Cambridge 36-inch telescope for another week. By Friday, 17th January, I had observed four more white dwarf stars and as it was a stormy night, I listed all those that I had observed in November or January and another eight 'highly desirable'.

Next day news came (via the grapevine) of the detection of optical flashes from the Crab Nebula, with mean brightness around 18th magnitude. During Sunday afternoon I carried out an order-of-magnitude calculation and estimated that the observation I had already made in November might reveal the flashes with a signal-to-noise ratio of about 6. Later the same day I observed the white dwarf van Maanen 1, and shortly before midnight the Crab Nebula, this time through the U, B, V filters as well as with no filter. The IAU Circular which reported Cocke, Disney and Taylor's (1969) optical discovery of the Crab pulsar did not arrive in Cambridge until 24th January.

Thanks to the Titan computer, I was able to reduce all of the Crab Nebula observations in a matter of days, searching only a small range of periods, and submitted a short paper to *Nature*, received there on 6th February and published on 15th March (Willstrop 1969a).

John Jelley returned to Cambridge for the week 10–17 February and we used a two-channel photometer with coincidence circuits to search for fine structure (~ 20 ns) in the flashes of light from the Crab pulsar, but nothing was detected (Jelley & Willstrop 1969).

In March I took the photometer and some timing equipment from A.E.R.E. Harwell to Pretoria to observe the Vela pulsar PSR 0833 -45. A multi-channel scaler was provided on loan from the South African Atomic Energy Board. The uncertainty in the position of this pulsar at that time set a limit to the integration time that could be used on each 6×6 arcsec square of about 3 minutes (2000 cycles = 178.4 sec). Nothing was found, down to a time-averaged magnitude B = 22 (Willstrop 1969b). Several years later, a group led by F. Graham Smith used the Anglo-Australian Telescope and found flashes with an average magnitude of B = 24.5. Graham was kind enough to phone to tell me of their success shortly before their result was published because he remembered that I had made an earlier search.

My colleagues were sympathetic at all times, and several people outside the Department provided help: in December 1968 Chris Cheney of the Computer Laboratory wrote some machine-code instructions to pack data from the paper tapes more densely in the Titan computer, to allow longer runs to be analyzed, and Clive Page from Radio Astronomy supplied a copy of the Cooley-Tukey fast Fourier transform in February 1969, which made it possible to analyze 16384 (2^{14}) readings for all 8192 possible frequencies in less than a minute. The Astronomer Royal, Sir Richard Woolley, transferred some nights on the I.N.T. originally allocated to the R.G.O. to my use, in addition to time granted by the Large Telescope Users' Panel. Ed Kibblewhite collaborated with me in writing a detailed description (Kibblewhite & Willstrop 1971) of the photometer and all the observations made at Cambridge and Herstmonceux. The support of R. O. Redman, Head of the Department, was vital to obtain time on the Radcliffe reflector and the I.N.T. His wry comment was meant to be, and was accepted as, sympathetic: 'These things are apt to be irritating to a sensitive skin.'

A Close Call

The reasons that I did not analyze the Crab Nebula data from 24/25 November any earlier, i.e. before Cocke and Disney's discovery were:

1. The Crab Nebula was the only object which I had observed at the maximum time resolution (97.6 readings/second) and at the time of the observation Ed Kibblewhite and I were unsure whether the recording equipment was operating reliably at that rate. It was only later that we verified that it was reliable.

2. There had been five nights in a week in which it was possible to obtain some observations, and I had accumulated a stack of data on white dwarf stars. Hewish *et al.* had postulated that pulsations of white dwarf stars were the origin of the radio pulses first detected from CP1919 etc., and one of my aims was to show whether or not white dwarf stars showed fast optical pulsations. They don't.

3. At the time of my observations in November 1968 the position of the fast pulsar was still uncertain by 10 arc minutes, which could have put it well outside the Crab Nebula. My photometer had a search field of view 5 arc minutes in diameter, but that would have admitted so much light from the sky that it would have masked any small fluctuations in the object itself. As it was, with the photometer diaphragm subtending 50 arc seconds on the

sky, the fluctuation I eventually found was of the order of 1 per cent of the total light flux.

4. Shortage of computer time, and an inefficient form of Fourier Transform program, written by myself. I was using all of my allocation of time on the Titan computer in analyzing the white dwarf data. Don't forget that all this was 38 years ago when computer speeds and data storage capacity were microscopic by modern standards. My first analysis of the Crab data, after Cocke and Disney's discovery, covered only a small range of pulse frequencies. Of course, this could have been done earlier, and with 20/20 hindsight it is obvious that that is what I should have done, but no previous searches for optical flashes had been successful.

References

Cocke, W.J., M.J. Disney & D.J. Taylor, (1969) I.A.U. Circular 2128.

Hewish, A., S.J. Bell, J.D.H. Pilkington, P.F. Scott & R.A. Collins, (1968) Nature, 217, 709.

Jelley, J.V. & R.V. Willstrop, (1968) Nature, 218, 753.

Jelley, J.V. & R.V. Willstrop, (1969) Nature, 224, 568.

Kibblewhite, E.J., & R.V. Willstrop, (1971) Mon. Not. R. astr. Soc., 154, 301.

Lovelace, R.B.E., J.M. Sutton & H.D. Craft, (1968) I.A.U. Circular 2113.

Staelin, D.H., & E.C. Reifenstein, (1968) I.A.U. Circular 2110.

Steavenson, W.H., (1938) Mon. Not. R. astr. Soc., 98, 673.

Steavenson, W.H., (1939) Mon. Not. R. astr. Soc., 99, 697.

Steavenson, W.H., (1946) Mon. Not. R. astr. Soc., 106, 280.

Steavenson, W.H., (1947) Mon. Not. R. astr. Soc., 107, 401.

Steavenson, W.H., (1948) Mon. Not. R. astr. Soc., 108, 186.

Steavenson, W.H., (1950) Mon. Not. R. astr. Soc., 110, 621.

Steavenson, W.H., (1953) Mon. Not. R. astr. Soc., 113, 258.

Walker, M.F., (1954) Publ. astr. Soc. Pacific, 66, 230.

Willstrop, R.V., (1969a) Nature, 221, 1023.

Willstrop, R.V., (1969b) Nature, 223, 281.

Dr Roderick Willstrop is a retired astronomer from the Institute of Astronomy at Cambridge University in England. He conducted the first observations that recorded optical pulses from a pulsar (specifically, the Crab pulsar in 1968). Unfortunately, as described in Chapter 7, the detection went unnoticed until after the famous discovery by Cocke, Disney and Taylor. Dr Willstrop once told the author, 'Looking back on the episode, I can see that my own activity was more of the "frantic, but in wrong directions" rather than "indolence" (which was my initial feeling at the time).'

Index